The Intelligent Universe

by James Gardner

AI, ET, and the Emerging Mind of the Cosmos

New Page Books
A division of The Career Press, Inc.
Franklin Lakes, NJ

THE INTELLIGENT UNIVERSE
EDITED AND TYPESET BY ASTRID DERIDDER
Cover design by Dutton and Sherman
Printed in the U.S.A. by Book-mart Press

To order this title, please call toll-free 1-800-CAREER-1 (NJ and Canada: 201-848-0310) to order using VISA or MasterCard, or for further information on books from Career Press.

CAREER PRESS

New Page BOOKS

The Career Press, Inc., 3 Tice Road, PO Box 687,
Franklin Lakes, NJ 07417
www.careerpress.com
www.newpagebooks.com

Library of Congress Cataloging-in-Publication Data

Gardner, James N.
Intelligent universe : AI, ET, and the emerging mind of the cosmos / by James Gardner.
p. cm.
Includes index.
ISBN-13: 978-1-56414-919-0
ISBN-10: 1-56414-919-6
1. Exobiology. 2. Extraterrestrial anthropology. 3. Life on other planets. 4. Human-alien encounters. 5. Fermi's paradox. 6. End of the universe. I. Title.

QH327.G37 2007
576.8'39--dc22

2006030552

The Intelligent Universe

Epigraph

Looking back from the year 3000—some 40 generations hence—most of the historical and political issues that concern us now will have been forgotten. World War II will seem as distant as the Battle of Hastings does to us now. The geopolitical landscape will have transformed into an astropolitical landscape. Our science will seem quaint and embryonic. However, the desire to better know our place in the universe, to push the frontiers, to explore beyond one more barrier, will remain.

—Steven J. Dick, chief historian, NASA

Dedication

This book is dedicated to the memory of Fred Hoyle, author of the first *The Intelligent Universe*, whose scientific imagination soared far beyond the bonds of his home planet and who sought to touch the face of that unknown superintelligence, whose fingerprints he clearly perceived in the mysteriously life-friendly details of stellar nucleosynthesis.

Fred Hoyle's intellectual style and blunt manner did not always endear him to the scientific establishment. As Hoyle's biographer Simon Mitton notes, although stellar evolution was the common theme of the 1983 Nobel award for physics, Hoyle was suspiciously denied a share of the prize. Some astronomers, Mitton observes, "still feel that the [Nobel] academy committed a gross injustice by declining to split the award three ways, thus including Hoyle."[1]

If the proponents of the many-worlds interpretation of quantum physics are correct and there are multiple versions of reality, some of which differ only slightly from our own cosmos, then I would like to imagine that in one of those worlds—a more just and fair-minded world, to be sure, than that which we inhabit—there is a newspaper archive somewhere containing an old, yellowed copy of a 1983 issue of *The New York Times* announcing that Sir Fred Hoyle, astronomer and astrophysicist extraordinaire, has just been awarded the Nobel Prize in Physics for his groundbreaking contributions to humanity's understanding of the nucleosynthetic process by which the chemical elements so utterly essential to life are forged in the hearts of giant supernova conflagrations.

The Intelligent Universe

Acknowledgments

Many people helped me in conceiving and writing this book. So a tip of the hat, in no particular order, to the following:

Natasha Kern, my literary agent, who possesses amazing reservoirs of patience, skill, and tenacity.

All the great people at Career Press/New Page Books, especially Michael Pye and my editor, Astrid deRidder.

The distinguished scientists, academicians, and futurists (especially Ray Kurzweil, Freeman Dyson, Seth Shostak, John Casti, Simon Mitton, and Steven J. Dick) who read the manuscript (or the essays on which it is based) and offered their insights and critiques. Any errors that remain are solely my responsibility.

My sweet and smart wife Lynda—truly, in the words of songsters Jesse Belvin and Curtis Williams, an Earth angel—who has an awesome capacity for love and support.

And, finally, to the beautiful and mysterious universe itself—veiled cosmic dancer inseparable from the dance of life—that is both as distant as the farthest star and as near as the nose on your face.

The Intelligent Universe

Contents

The Intelligent Universe

Foreword

by Ray Kurzweil

Consider that the price-performance of computation has grown at a super-exponential rate for over a century. The doubling time (of computes per dollar) was three years in 1900 and two years in the middle of the 20th century; and price-performance is now doubling each year. This progression has been remarkably smooth and predictable through five paradigms of computing substrate: electromechanical calculators, relay-based computers, vacuum tubes, transistors, and now several decades of Moore's Law (which is based on shrinking the size of key features on a flat integrated circuit). The sixth paradigm—three-dimensional molecular computing—is already beginning to work and is waiting in the wings. We see similar smooth exponential progressions in every other aspect of information technology, a phenomenon I call the law of accelerating returns.

Where is all this headed? It is leading inexorably to the intelligent universe that Jim Gardner envisions. Consider the following: As with all of the other manifestations of information technology, we are also making exponential gains in reverse-engineering the human brain. The spatial resolution in 3D volume of in-vivo brain scanning is doubling each year, and the latest generation of scanners is capable of imaging individual interneuronal connections and seeing them interact in real time. For the first time, we can see the brain create our thoughts, and also see our thoughts create our brain (that is, we create new spines and synapses as we learn). The amount of data we are gathering about the brain is doubling each year, and we are showing that we can turn this data into working models and simulations.

Already, about 20 regions of the human brain have been modeled and simulated. We can then apply tests to the simulations and compare these results to the performance of the actual human brain regions. These tests have had impressive results, including one of a simulation of the cerebellum, the region responsible for physical skill, and which comprises about half of the neurons in the brain. I make the case in my book (*The Singularity is Near*) that we will have models and simulations of all several hundred regions, including the cerebral cortex, within 20 years. Already, IBM is building a detailed simulation of a substantial portion of the cerebral cortex. The result of this activity will be greater insight into ourselves, as well as a dramatic expansion of the AI tool kit to incorporate all of the methods of human intelligence.

By 2029, sufficient computation to simulate the entire human brain, which I estimate at about 10^{16} (10 million billion) calculations per second (cps), will cost about a dollar. By that time, intelligent machines will combine the subtle and supple skills that humans now excel in (essentially our powers of pattern recognition) with ways in which machines are already superior, such as remembering trillions of facts accurately, searching quickly through vast databases, and downloading skills and knowledge.

But this will not be an alien invasion of intelligent machines. It will be an expression of our own civilization, as we have always used our technology to extend our physical and mental reach. We will merge with this technology by sending intelligent nanobots (blood-cell-sized computerized robots) into our brains through the capillaries to intimately interact with our biological neurons. If this scenario sounds very futuristic, I would point out that we already have blood-cell-sized devices that are performing sophisticated therapeutic functions in animals, such as curing Type I diabetes and identifying and destroying cancer cells. We already have a pea-sized device approved for human use that can be placed in patients' brains to replace the biological neurons destroyed by Parkinson's disease, the latest generation of which allows you to download new software to your neural implant from outside the patient.

If you consider what machines are already capable of, and apply a billion-fold increase in price-performance and capacity of computational technology over the next quarter century (while at the same time we shrink the key features of both electronic and mechanical technology by a factor of 100,000), you will get some idea of what will be feasible in 25 years.

By the mid-2040s, the nonbiological portion of the intelligence of our human-machine civilization will be about a billion times greater than the biological portion (we have about 10^{26} cps among all human brains today; nonbiological intelligence in 2045 will provide about 10^{35} cps). Keep in mind that, as this happens, our civilization will be become capable of performing more ambitious engineering projects. One of these projects will be to keep this exponential growth of computation going. Another will be to continually redesign the source code of our own intelligence. We cannot easily redesign human intelligence today, given that our biological intelligence is largely hard-wired. But our future—largely nonbiological—intelligence will be able to apply its own intelligence to redesign its own algorithms.

So what are the limits of computation? I show in my book that the ultimate one-kilogram computer (less than the weight of a typical notebook computer today) could perform about 10^{42} cps if we want to keep the device cool, and about 10^{50} cps if we allow it to get hot. By hot, I mean the temperature of a hydrogen bomb going off, so we are likely to asymptote to a figure just short of 10^{50} cps. Consider, however, that by the time we get to 10^{42} cps per kilogram of matter, our civilization will possess a vast amount of intelligent engineering capability to figure out how to get to 10^{43} cps, and then 10^{44} cps, and so on.

So what happens then? Once we saturate the ability of matter and energy to support computation, continuing the ongoing expansion of human intelligence and knowledge (which I see as the overall mission of our human-machine civilization), will require converting more and more matter into this ultimate computing substrate, sometimes referred to as "computronium."

What is that limit? The overall solar system, which is dominated by the sun, has a mass of about 2×10^{30} kilograms. If we apply our 10^{50} cps per kilogram limit to this figure, we get a crude estimate of 10^{80} cps for the computational capacity of our solar system. There are some practical considerations here, in that we won't want to convert the entire solar system into computronium, and some of it is not suitable for this purpose anyway. If we devoted 1/20th of 1 percent (.0005) of the matter of the solar system to computronium, we get capacities of 10^{69} cps for "cold" computing and 10^{77} cps for "hot" computing. I show in my book how we will get to these levels using the resources in our solar system within about a century.

I'd say that's pretty rapid progress. Consider that in 1850, a state-of-the-art method to transmit messages was the Pony Express, and calculations were performed with an ink stylus on paper. Only 250 years later, we will have vastly expanded the intelligence of our civilization. Just taking the 10^{69} cps figure, if we compare that to the 10^{26} cps figure, which represents the capacity of all human biological intelligence today, that will represent an expansion by a factor of 10^{43} (10 million trillion trillion trillion).

Now for the intelligent universe. At this point, the ongoing expansion of our intelligence will require moving out into the rest of the universe. Indeed, this process will start before we saturate the resources in our midst. When this happens, we will immediately confront a key issue—the speed of light—which we understand to be the cosmic speed limit. But what is it a speed limit for? We can easily cite examples of phenomena that exceed the speed of light. For example, we know the universe to be expanding, and the speed with which galaxies recede from each other exceeds the speed of light if the distance between the two galaxies is greater than what is called the Hubble distance.

But the speed of light, as postulated by Einstein in his special theory of relativity, represents a limit on the speed with which we can transmit information. The phenomenon of receding galaxies does not violate Einstein's theory because it is caused by space expanding, rather than the galaxies moving through space. As such, it does not help us to transmit information at speeds faster than the speed of light.

Another phenomenon that appears to exceed the speed of light is quantum disentanglement of two entangled particles. Two particles created together may be "quantum entangled," meaning that if we resolve the ambiguity of a undetermined property (such as the phase of its spin) in one of the paired particles (by measuring it), it will also be resolved in the other particle as the same value, and at exactly the same time. There is the appearance of some sort of communication link between the two particles, and this phenomenon has been experimentally measured at many times the speed of light. But again, this does not allow us to transmit information (such as a file), because what is being "communicated" by quantum disentanglement is not information, but quantum randomness. As such, it can be used to generate profoundly random encryption codes (and that application has already been exploited in a new generation of quantum encryption devices), but it does not allow faster-than-light communication.

There are suggestions that the speed of light has changed slightly. In 2001, astronomer John Webb presented results that suggested that the speed of light may have changed by 4.5 parts out of 10^8 over the past 2 billion years. These observations need confirmation. That may not seem like much of a change, but it is the nature of engineering to take a subtle effect and amplify it. So perhaps there are ways to engineer a change in the speed of light.

The theory that the early universe went through a rapid expansion in an inflationary period does postulate a speed far greater than the speed of light, so we may be able to find an engineering approach to harnesses the conditions that existed in the early universe.

The most compelling idea of circumventing the speed of light is not to change it at all, but simply to find shortcuts to places in the universe that seem to be far away. The theory of general relativity does not rule out the existence of wormholes in time-space that could allow us to travel to a far-off location in a short period of time. California Institute of Technology physicists Michael Morris, Kip Thorne, and Uri Yurtsever have described theoretical methods to engineer wormholes to get to far-away locations in a brief period of time. The amount of energy required might make it difficult to set up a passageway for biological humans to pass through, but our exploration and colonization of the universe requires only nanobots.

Physicists David Hochberg and Thomas Kephart have shown how gravity was strong enough in the very early universe to have provided the energy required to spontaneously create massive numbers of self-stabilizing wormholes. A significant portion of these wormholes is likely to still be around and may be pervasive, providing a vast network of corridors that reach far and wide throughout the universe. It might be easier to discover and use these natural wormholes than to create new ones.

We have to regard these proposals to exceed or bypass the speed of light as speculative. But while this may be regarded as an interesting intellectual reflection today, it will be the primary issue confronting human civilization a century from now. And keep in mind that we're talking about a civilization that will be trillions of trillions of times more capable than we are today. So one thing we can be confident of, is that if there is any way to transmit devices and information at speeds exceeding the speed of light (or circumventing it through wormholes), our future civilization will be both motivated and capable of discovering and exploiting that insight.

The price-performance of computation went from 10^{-5} to 10^8 cps per thousand dollars in the 20th century. We also went from about a million dollars to a trillion dollars in the amount of capital devoted to computation, so overall progress in nonbiological intelligence went from 10^{-2} to 10^{17} cps in the 20th century, which is still short of the human biological figure of 10^{26} cps. We will achieve around 10^{69} cps by the end of the 21st century. If we can circumvent the speed of light, we only need about another 20 orders of magnitude to convert the entire universe into computronium, and that can be done well within another century. On the other hand, if the speed of light remains unperturbed by the vast intelligence that will seek to overcome it, it will take billions of years. But it will still happen.

I make this case more fully in my book, and Jim makes it quite forcefully in this book. It is remarkable to me that almost all of the discussions of cosmology fail to mention the role of intelligence. In the common cosmological view, intelligence is just a bit of froth, something interesting that happens on the sidelines of the great cosmic story. But in the standard view, whether the universe winds up or down, ends up in fire (a great crunch and new Big Bang), or ice (an ever-expanding and ultimately dead universe), or something in-between, depends only on measures of dark matter, dark energy, and other parameters we have yet to discover. That the story of the universe is a story yet to be written by the intelligence it will spawn is almost never mentioned. This book will help to change the common "unintelligent" view.

So what will we do when our intelligence is in the range of a google (10^{100}) cps? One thing we may do is to engineer new universes. Similarly, our universe may be the creation of some superintelligences in another universe. In this case, there was an intelligent designer of our universe—that designer would be the evolved intelligence of some other universe that created ours. Perhaps our universe is a science fair experiment of a student in another universe. (Reading the news of the day, you might get the impression that this erstwhile adolescent superintelligence who designed our universe is not going to get a very good grade on his or her project.)

But the evolution of intelligence here on Earth is actually going very well. All of the vagaries (and tragedies) of human history, such as two world wars, the cold war, the great depression, and other notable events, did not make even the slightest dent in the ongoing exponential progressions I previously mentioned.

Clearly, the universe we live in does appear to be an intelligent design, in that the constants in nature are precisely what are required for the universe to have grown in complexity. If the cosmological constant, the Planck constant, and the many other constants of physics were set to just slightly different values, atoms, molecules, stars, planets, organisms, humans, and this book would have been impossible. As Jim Gardner says, "A multitude of...factors are fine-tuned with fantastic exactitude to a degree that renders the cosmos almost spookily bio-friendly." How the rules of the universe happened to be just so is a profound question, one that Gardner explores in fascinating detail.

Or perhaps our universe is not someone's science experiment, but rather the result of an evolutionary process. Leonard Susskind, the developer of string theory, and Lee Smolin, a theoretical physicist and expert on quantum gravity, have suggested that universes give rise to other universes in a natural, evolutionary process that gradually refines the natural constants. Smolin postulates that universes best able to product black holes are the ones that are most likely to reproduce. Smolin explains, "Reproduction through black holes leads to a multiverse in which the conditions for life are common—essentially because some of the conditions life requires, such as plentiful carbon, also boost the formation of stars massive enough to become black holes."[1]

As an alternative to Smolin's concept of it being a coincidence that black holes and biological life both need similar conditions (such as large amounts of carbon), Jim Gardner and I have put forth the conjecture that it is precisely the intelligence that derives from biological life and its technological creations that are likely to engineer new universes with intelligently set parameters. In this thesis, there is still an important role for black holes, because black holes represent the ultimate computer. Now that Stephen Hawking has conceded that we can get information out of a black hole (because the particles comprising the Hawking radiation remain quantum-entangled with particles flying into the black hole), the extreme density of matter and energy in a black hole make it the ultimate computer. If we think of evolving universes as the ultimate evolutionary algorithm, the utility function (that is, the property being optimized in an evolutionary process) would be its ability to produce intelligent computation.

This line of reasoning sheds some light on the Fermi paradox. The Drake formula provides a means to estimate the number of intelligent civilizations in a galaxy or in the universe. Essentially, the likelihood of a planet evolving biological life that has created sophisticated technology is tiny, but there are so many star systems, that there should still be many millions of such civilizations. Carl Sagan's analysis of the Drake formula concludes that there should be around a million civilizations with advanced technology in our galaxy, while Frank Drake estimated around 10,000. And there are many billions of galaxies. Yet we don't notice any of these intelligent civilizations, hence the paradox that Fermi described in his famous comment. As Jim Gardner and others have asked, where is everyone?

We can readily explain why any one of these civilizations might be quiet. Perhaps it destroyed itself. Perhaps it is following the *Star Trek* ethical guideline to avoid interference with primitive civilizations (such as ours). These explanations make sense for any one civilization, but it is not credible, in my view, that every one of the billions of technology capable civilizations that should exist has destroyed itself or decided to remain quiet.

The SETI project is sometimes described as trying to find a needle (evidence of a technical civilization) in a haystack (all the natural signals in the universe). But actually, any technologically sophisticated civilization would be generating trillions of trillions of needles (noticeably intelligent signals). Even if they have switched away from electromagnetic transmissions as a primary form of communication, there would still be vast artifacts of electromagnetic phenomenon generated by all of the many computational and communication processes that such a civilization would need to engage in.

Now let's factor in the law of accelerating returns. The common wisdom (based on what I call the intuitive linear perspective) is that it would take many thousands, if not millions of years, for an early technological civilization to become capable of technology that spanned a solar system. But as I argued previously, because of the explosive nature of exponential growth, it will only take a quarter of a millennium (in our own case) to go from sending messages on horseback to saturating the matter and energy in our solar system with sublimely intelligent processes.

According to most analyses of the Drake equation, there should be billions of civilizations, and a substantial fraction of these should be ahead of us by millions of years. That's enough time for many of them to be capable of vast galaxy-wide technologies. So how can it be that we haven't noticed any of the trillions of trillions of "needles" that each of these billions of advanced civilizations should be creating?

My own conclusion is that they don't exist. If it seems unlikely that we would be in the lead in the universe, here on the third planet of a humble star in an otherwise undistinguished galaxy, it's no more perplexing than the existence of our universe with its ever so precisely tuned formulas to allow life to evolve in the first place.

It is not possible to do justice to this dilemma in a foreword. It would take a book to do that, and Jim Gardner has written that book. Muriel Rukeyser wrote, "The universe is made of stories, not atoms," and in this book, Gardner tells us the universe's own fascinating and unfinished story. Perhaps even more intriguing, Gardner relays in a clear and compelling manner the gripping stories of the rich, intellectual ferment from which our understanding of the universe is emerging.

Introduction

It Takes a Giant Cosmos to Create Life and Mind

There is a time machine clearly visible right outside your front door. It's easy to see—in fact, it's impossible to overlook—although its awesome powers are generally ignored by all but a discerning few. The unearthly beauty, the ineffable grandeur, and the ingenuity of construction of this time machine are humbling to every human being who makes an effort to probe into the enigma of its origin and the mystery of its ultimate destiny. The time machine of which I speak is emphatically not of human origin. Indeed, a few venturesome scientists are beginning to entertain a truly incredible possibility: that this device is an artifact bequeathed to us by a supreme intelligence that existed long, long ago and far, far away. All knowledgeable observers agree that the scope of its stupendous powers and the sheer delicacy of its miniscule moving parts seem nothing short of miraculous.

A second amazing but incontrovertible fact confronts those trained in the science of cosmology: We human beings are living our daily lives in the midst of extraterrestrial entities. These entities are everywhere—in the air we breathe, in the food we eat, in the ground beneath our feet, and inside our bodies. These extraterrestrials have made an incredible journey from the venue of their birth to reach planet Earth. Their epic migration, spanning millions of light-years, dwarfs the fictional interstellar voyages of the starship *Enterprise*. They are the real star trekkers, with more mileage on their odometers than we are capable of imagining. And perhaps most astonishing, we could not possibly survive without their constant presence, and the unfailing exercise of their special powers.

Could the existence of this purported time machine be anything but outrageous science fiction? And how could there be extraterrestrials among us that we have never noticed? Surely not even an inebriated television producer would find these ideas sufficiently credible to weave into an *X-Files* plot!

Yet I can assure you that both propositions are correct. Indeed, they are indisputable.

The time machine is the universe itself. We see its local features every night in the starry sky above us. The firmament we observe is not a picture of the stars and galaxies as they exist today, but rather a kind of cinematic image of our corner of the cosmos as it existed years ago—in the case of the great galaxy Andromeda, millions of years ago. Because starlight travels through the immensity of interstellar and intergalactic space at a finite pace, and because of the inconceivable vastness of the cosmos, we look backward in time with every glance at the nighttime sky.

With powerful spectacles to aid our vision—massive instruments such as the telescopes that dot the peak of Mauna Kea in Hawaii and the Hubble Space Telescope—we can extend our gaze incredibly far back into the past, indeed virtually to the moment of the Big Bang. And with even more sophisticated observational instruments, such as the Advanced Laser Interferometer Gravitational-Wave Observatory (LIGO) and the space-based Big Bang Observer (BBO) that NASA hopes to deploy by 2025, there is hope that we will be able to glimpse the moment of cosmic creation itself—the very genesis of space and time.

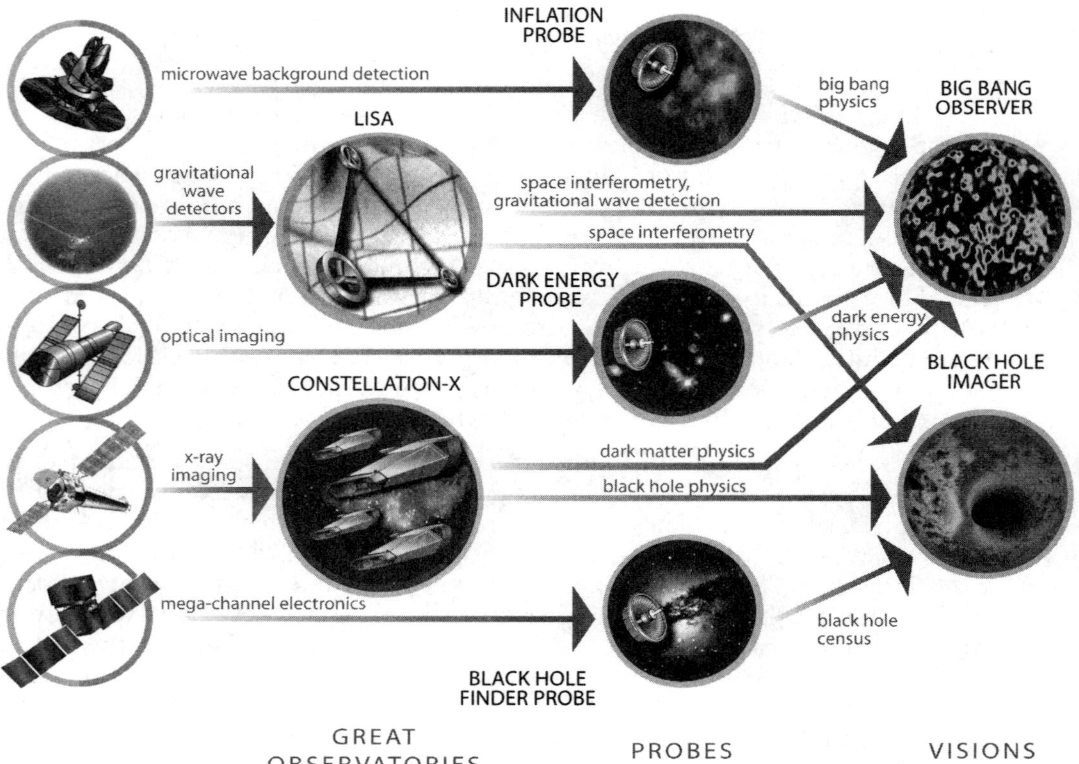

NASA flowchart of "Beyond Einstein" missions
Image provided by NASA

What about those extraterrestrials? They are the atoms that combine to form the molecules from which our bodies and virtually everything else in our world and the solar system are made. These extraterrestrials were not, for the most part, born *ex nihilo* in the fireball of the Big Bang. Instead, they were hammered into existence in the forges of supernova explosions—rare conflagrations that release more energy in a flash than the normal output of the billions of ordinary stars in a typical galaxy.

Of all these extraterrestrial entities, the one with the most unusual birth story is carbon, the essential foundation of life as we know it. The peculiar process of stellar alchemy by which elemental carbon is coaxed into existence is so delicate and improbable that it prompted a giant of British astronomy, Sir Fred Hoyle, to utter the most famous and controversial remark of his storied career:

> Would you not say to yourself, "Some super-calculating intellect must have designed the properties of the carbon atom, otherwise the chance of my finding such an atom through the blind forces of nature would be utterly minuscule?" Of course you would.... A common sense interpretation of the facts suggests that a superintellect has monkeyed with physics, as well as with chemistry and biology, and that there are no blind forces worth speaking about in nature. The numbers one calculates from the facts seem to me so overwhelming as to put this conclusion almost beyond question.[1]

Hoyle's remark is the inspiration for *The Intelligent Universe*. The book is the story of an idea, and the idea is quite simple: The best way to think about life, intelligence, and the universe is that they are not separate things, but are different aspects of a single phenomenon. To take liberties with a popular ballad, "We are the world, we are the people, *and* we are the universe." To state this proposition from the opposite perspective, the universe is coming to life and waking up through the processes of our lives and thoughts, and, very probably, through the lives and thoughts of countless other beings scattered throughout the cosmos.

One startling implication of this idea is that the *true* story of the origin of the human species is longer than the saga of terrestrial evolution conceived of by Charles Darwin and his intellectual progeny. Thanks to the discoveries of Hoyle and other cosmologists, it is now beyond dispute that the life history of humanity includes the entire history of the cosmos itself. Why? Because an inconceivably ancient and immense universe is needed to create even one species of minuscule living creatures on a single planet orbiting a nondescript star in the outer reaches of an ordinary galaxy.

If the cosmos were not so old and large, multiple generations of stars could not have formed, burned brightly for billions of years, and then blown themselves to pieces in titanic supernovae explosions, thereby synthesizing all the higher elements in the periodic table. Absent those elements (especially carbon and oxygen), there could be no life anywhere amid the countless galaxies that fill the universe.

A second implication of this concept is that if extraterrestrial life and intelligence should exist, it will inevitably be related to mankind. No, I am not talking about a government-suppressed history of alien visitation and cross-breeding, or even the slightly more plausible scenario outlined by Nobel laureate Francis Crick of *directed panspermia*.

Directed Panspermia

In *Life Itself: Its Origin and Nature*[2] Nobel laureate Francis Crick, co-discoverer of the double helix structure of DNA, put forward a hypothesis about the origin of life on Earth that many of his scientific colleagues viewed as outlandish, even scandalous. The essence of Crick's scenario was that, contrary to Darwin's speculation that the first living things may have emerged spontaneously in a warm little pond, terrestrial life was deliberately seeded by an advanced alien race billions of years ago. Crick's ideas built on those of Swedish physicist Svante August Arrhenius, who suggested in the late 19th century that life did not get started on Earth, but was seeded by microorganisms drifting in from outer space under the gentle pressure of ambient starlight.

A perceived weakness of Arrhenius's theory—called simply *panspermia*, which translates literally as *seeds everywhere*—was that it was thought unlikely that spores or microorganisms could survive the harsh radiation of space for the decades, centuries, or even millennia that would be required for bacteria to slowly waft from even the nearest stars to our solar system.

Crick sought to remedy this weakness in Arrhenius's theory by proposing that the transplanted extraterrestrial microorganisms had actually traveled to Earth within the protective hull of an alien spaceship! As Crick put it:

> Life started here when these organisms were dropped into the primitive ocean and began to multiply.[3]

Why would this obviously serious-minded and gifted scientist put forward such a seemingly eccentric proposal? Essentially, Crick was attempting to take seriously the logical implications of what he recognized as "the very high degree of [the] *organized complexity* [of living things] we find at every level, and especially at the molecular level."[4] In order for even the simplest living creature to metabolize and reproduce, a vast array of incredibly complicated and interdependent molecular machinery must function, at a nanoscale level, with a degree of flawless precision that makes the operations of a Boeing 747 look downright primitive by comparison. As Crick put it in a candid and colorful remark that has become a key talking point for the Intelligent Design crowd:

> The origin of life appears at the moment to be almost a miracle, so many are the conditions which would have had to have been satisfied to get it going. [5]

But if life originated on an alien world and was later transported here by a race of intelligent aliens, then the probabilistic resources available to explain a random origin of life's organized complexity can be expanded exponentially. The major conceptual weakness of Crick's directed panspermia scenario is that it merely postpones the ultimate question: How did life originally get going, either on a distant planet or in that proverbial warm little pond right here on Earth?

I am asserting that wherever and however life and intelligence may exist elsewhere in the cosmos, it will have originated and evolved from a universally shared substrate: the chemical elements of the periodic table and the basic forces and parameters of physics. As far as anyone can tell, these elements, forces, and parameters appear invariant throughout the visible universe. They can be thought of as a kind of "deep DNA"—a universal genetic code inscribed far below the level of terrestrial genomes. At this fundamental level, everyone and everything that exists in the universe, whether animate or inanimate, is intimately related. And because all of this living and not-yet-living stuff owes its ultimate origin to a common genesis event (the Big Bang), we are all related in a family way. With apologies to Saint Francis of Assisi, we can confidently state that Earth's satellite truly is Sister Moon, and that the life-giving star 93 million miles away is genuinely Brother Sun.

A third implication of the concept is that because the vast preponderance of the lifetime of the universe lies in the distant future rather than in the past, the historical achievements of life and mind are meager foreshadowings of the starring role that intelligent life is likely to play in shaping the future of the cosmos. Indeed, this new way of looking at the intimate linkage of life, mind, and the cosmos suggests a novel way of thinking about the ultimate destiny of our universe.

Traditionally, scientists have offered two bleak answers to the profound issue of how the universe will end: *fire* or *ice*. The cosmos might end in *fire*—a cataclysmic Big Crunch in which galaxies, planets, and any life forms that might have endured to the end time are consumed in a raging inferno as the universe contracts in a kind of Big Bang, but in reverse.

Or the universe might end in *ice*—a ceaseless expansion of the fabric of space-time in which the thin soup of matter and energy is eternally diluted and cooled. Under this scenario, stars wither and die, constellations of cold matter recede further and further from one another, and the vast project of cosmic evolution simply fades into quiet and endless oblivion.

The Intelligent Universe proposes a third possibility: that the universe might end in *intelligent life*. Not life as we know it, but life that has acquired the capacity to shape the cosmos as a whole, just as life on Earth has acquired the ability to shape the land, the sea, and the atmosphere. As Princeton physicist Freeman Dyson puts it:

> Mind, through the long course of biological evolution, has established itself as a moving force in our little corner of the universe. Here on this small planet, mind has infiltrated matter and has taken control. It appears to me that the tendency of mind to infiltrate and control matter is a law of nature.[6]

My first book, *Biocosm*,[7] was one long argument that the cosmos possesses a utility function (some value or outcome that is being maximized) and that the specific utility function of our cosmos is propagation of baby universes exhibiting the same life-friendly physical qualities as their parent-universe. Under this scenario, the mission of sufficiently evolved intelligent life in the universe is essentially to serve as a cosmic reproductive organ, spawning an endless succession of life-friendly offspring that are themselves endowed with the same reproductive capacities as their predecessors. The fact that our universe seems queerly hospitable to carbon-based intelligent life—an astronomically improbable oddity that many leading scientists have identified as the deepest mystery in all of science—emerges in the context of this hypothesis as a predictable outcome (a *falsifiable retrodiction*, in the jargon of science).

Falsifiable Retrodictions

Traditionally, scientists insist that new hypotheses generate falsifiable *predictions* of experimental results in order to qualify as genuine science. However, there are some fields of science—especially archaeology and cosmology, which involve events that occurred in the distant past or in physically inaccessible regions—that cannot generate predictions susceptible to laboratory testing. Although a few purists regard these fields as intrinsically unscientific, most scientists concede that it is appropriate for so-called "historical" sciences, such as geology, evolutionary biology, cosmology, paleontology, and archaeology to rely on retrodiction as an alternate means of testing a scientific hypothesis. A retrodiction essentially compares previously gathered observational evidence (for instance, the fossil record, in the case of evolutionary biology) with the implications of a scientific hypothesis (such as Darwinian natural selection). If the observational evidence agrees with the implications of the hypothesis, the hypothesis is said to retrodict the evidence. A detailed discussion of retrodiction as a tool for testing scientific hypothesis is contained in Appendix A.

Though *The Intelligent Universe* reprises some of the key themes of *Biocosm*, its primary objective is different. Unlike *Biocosm*, the purpose of this book is not to lay out a scientific hypothesis but rather to tell an extraordinary story—the story of the probable future of the universe. In telling this story, I am going to introduce you to some very unusual and interesting people.

You will meet a senior NASA official whose passion is investigating the probable impact on religion of the discovery of extraterrestrial intelligence. You will encounter a computer scientist who is coaxing software to undergo a special kind of Darwinian evolution, thus becoming more adept and financially valuable over time. And you will meet a technology prophet who, in my view, is the true contemporary heir to Darwin's intellectual legacy.

You will also meet a fascinating cast of nonhuman players likely to have leading roles on tomorrow's cosmic stage. They include: (1) super-smart machines capable of out-thinking humans without breaking a sweat; (2) speedy and cost-efficient interstellar probes that will consist of nothing more substantial than elaborate software algorithms capable of "living" in the innards of alien computers they may encounter on far-off planets; and (3) intelligent extraterrestrials, which SETI researchers have not yet discovered but whose probable existence is strongly predicted by my *Biocosm* hypothesis.

The Intelligent Universe, then, is a kind of projected travelogue—an imagined future history—of the cosmic journey that lies ahead. The foundation for that projection is a vision of the deep linkage between the three ostensibly separate phenomena previously mentioned: the appearance of life, the emergence of intelligence, and the seemingly mindless physical evolution of the cosmos. In discussing these topics, the book will not only provide news dispatches from the frontiers of cosmological science, but also offer musings about the philosophical implications of emerging scientific insights for our self-image as a species.

Some skeptics and traditionalists will doubtless protest that such philosophizing is out of place in a book that seeks to chronicle the latest scientific thinking about the nature of the universe. In rebuttal, I offer the timeless words of Galileo:

> Philosophy is written in this grand book—I mean the universe—which stands continually open to our gaze. But the book cannot be understood unless one first learns to comprehend the language and read the characters in which it is written.

In the spirit of Galileo, I invite you to gaze into this grand book—I mean our cosmos—and begin to learn the language and the characters in which it is written. As we shall see, the grand book is not only a tale of the past, but also a story about our tomorrows. Above all, it is a book that, carefully deciphered, foretells the incredible journey that intelligent life will make across the vast expanse of the cosmic future and the projected consummation of that voyage—the emergence of the biocosm.

Part I

Artificial Intelligence

This book begins with three chapters that focus on artificial intelligence, artificial life, and the intriguing suggestion by leading scientists that our cosmos may be a kind of giant natural computer running on what I call the Software of Everything. The Software of Everything consists of what the late physicist Heinz Pagels called the cosmic code—the full suite of physical constants and natural laws that prevail in our strangely and improbably life-friendly cosmos. The physical processors in the cosmic computer set-up are all the bits of matter and energy that fill the heavens.

At the heart of this arresting vision is an insistence on the cosmic primacy of information and computation. Strangely and miraculously, the vision reveals deep linkages and unanticipated correlations between the qualities of what we think of as inanimate nature and the essential characteristics of life itself.

Chapter 1

Hitching a Ride on the Software of Everything

It is hardly surprising that Galileo conceived of the cosmos as a book, open to all and waiting to be read. Printing and bookmaking, after all, were high technological accomplishments in his era—enablers of enlightenment and progenitors of the very concept of law in both natural and judicial contexts. Writing in the 1760s, the Italian jurist Cesare Becarria concluded that the most significant factor behind Europe's emergence from a dark age of lawless tyranny was not better rulers, better judges, or even better laws. It was rather "the art of printing, which makes the public, and not a few individuals, the guardians of the sacred laws."[1]

The art of printing played an equally critical role in catalyzing the birth of modern science. Galileo himself had been inspired by an early astronomical text—*On the Revolutions of the Celestial Spheres* by Nicolaus Copernicus, considered to be the father of modern astronomy. Galileo had been required to censor his personal copy of Copernicus's classic book in compliance with very specific instructions issued by the Vatican in 1620, though he made sure the deletions were sufficiently faint that the original text remained legible.[2] With respect to the heavenly book inscribed in the form of mathematical formulae tracking the movements of stars and planets, Galileo surely realized that no such strategy of evasion was required—the universal text was written in indelible ink, and no erasure or redaction would ever be possible, whatever Rome might decree.

Each era in history, it seems, has its own favorite metaphor with which to imaginatively conceive of the basic nature of the universe. Moreover, these metaphors seem to embody a key artistic or technological achievement of that era. For Galileo, the universe was a book, reflecting the crucial importance of printing and publishing to the dissemination of scientific theories and discoveries. In an earlier era, the Greeks conceived of the cosmos as a musical composition, reflecting the importance of music in their culture. For Isaac Newton, the cosmos was a vast clock, ticking off the moments of absolute time with the invariant precision that Einstein would subsequently show to be illusory.

The highest technological achievement of our own era is arguably software and computer fabrication, so it is not surprising that many leading thinkers have begun to speak of the universe as a kind of ultimate laptop, running on a mysteriously well-crafted operating system.

So let's consider the possibility that, contrary to Galileo's speculations, the universe is not actually a vast open book, but rather a massive natural computer. Its operating system would consist of the fundamental laws and dimensionless constants of physics. Its processing units would be the collage of stars, galaxies, planets, black holes, dust particles, molecules, and physical forces that populate the cosmos and shape its ongoing physical evolution. That's not as crazy as it sounds, as we'll see in a moment.

Stephen Wolfram and the Software of Everything

This seemingly bizarre conception of nature—the cosmos as a giant computer—is not as far-fetched as it might at first seem. In 2002, computer scientist and multimillionaire entrepreneur Stephen Wolfram self-published a massive tome that is indisputably the weightiest (literally and figuratively) popular science title. Roughly the size and weight of a large concrete building block, Wolfram's *A New Kind of Science* put forward an exceedingly simple—not to say lightweight—thesis. His basic idea, reiterated *ad nauseum,* was that:

> ...beneath all the complex phenomena we see in physics there lies some simple program which, if run for long enough, would reproduce our universe in every detail.[3]

Wolfram argued that the Holy Grail for which cosmologists should be searching is not a so-called Theory of Everything—a set of equations akin to $E = mc^2$ that might conceivably provide a unifying explanation for the diverse natural phenomena that are now the province of incommensurable scientific paradigms such as quantum physics and general relativity. Rather, Wolfram contends, scientists should be seeking a kind of Software of Everything—an elusive program that, if run long enough, would duplicate the entire cosmos and everything in it.

Wolfram's Software of Everything would be the mother of all source codes, generating not only the movements of stars and planets but also the emergence and evolution of life and intelligence:

> [W]ith such a program one would finally have a model of nature that was not in any sense an approximation or idealization. Instead, it would be a complete and precise representation of the actual operation of the universe—but all reduced to readily stated rules. In a sense, the existence of such a program would be the ultimate validation of the idea that

> human thought can comprehend the construction of the universe. But just knowing the underlying program does not mean that one can immediately deduce every aspect of how the universe will behave....[T]here is often a great distance between underlying rules and overall behavior. And in fact, this is precisely why it is conceivable that a simple program could reproduce all the complexity we see in physics.[4]

The idea may seem wacky, but Wolfram possesses unassailable credentials in the fields of science and technology. His impressive resume lists a Ph.D. at age 20 from Caltech, and a MacArthur Fellowship (often called a "genius grant") at age 21 (he's the youngest person ever to receive this prestigious honor). At age 23, Wolfram arrived at Princeton's Institute for Advanced Studies (Einstein's old haunt) with an audacious goal: to reduce the emergence of all of nature's complexity to the ceaseless operation of peculiar pattern-generating mechanisms called cellular automata.

Cellular automata are a special class of software programs that were discovered by the mathematician and computer pioneer John von Neumann. A cellular automaton is nothing but a simple computational mechanism that alters the state of a cell on a checkerboard-like grid—for instance, changing its color from white to black—depending on the state of other cells on the grid that are either adjacent to or near the target cell, in accordance with a simple transformational rule (such as the immediately prior colors of the target cell and its neighbors). The program is called an *automaton* because it blindly follows the simple rule with which it has been programmed over and over.

The truly astonishing thing about these simple little programs is their ability to generate a stunning degree of pattern diversity if allowed to operate long enough. Churning away mindlessly, cellular automata (CAs, for short) can disgorge intricate physical representations that bear an uncanny resemblance to complex and beautiful features of the natural world. Wolfram once inadvertently reproduced the stunning pattern of certain mollusk shells, for instance, just by running a simple little CA program repeatedly. He later discovered that other parallels to nature's own patterns began appearing like magic on his computer screen, including patterns indistinguishable from the geography of river basins and the intricate shape of snowflakes.

It turns out that this was no coincidence. It is now becoming clear that, at a deep level, the pattern-generating processes of nature operate very much like Wolfram's little cellular automata. Consider those mollusk shell patterns, for instance. They are generated by biological versions of exactly the same type of CA computer program that Wolfram was using in his laboratory. Nature, it seems, is doing something that looks suspiciously similar to computation.

Edward Fredkin and Digital Physics

Wolfram is not the only serious scientist who believes that the universe is a kind of vast natural computer with parallels to human-designed laptops. Computer guru Edward Fredkin famously advocates a vision of nature that he calls digital physics. The essence of Fredkin's vision, expressed in an audacious scientific paper, is his hypothesis that there will eventually be discovered a "single cellular automaton rule that models all of microscopic physics, and models it exactly."[5]

Fredkin, an eccentric multimillionaire entrepreneur, is a self-taught computer genius who views the world from a decidedly unusual perspective. But he's no crank. His intellectual skills are such that he was made a full professor at the Massachusetts Institute of Technology without ever having acquired a bachelor's degree. And his lifelong obsession with using the tools of computer science to illuminate the basic processes studied by physicists has yielded some unusual insights. His basic concept—that information is the basic stuff of which the universe is made, and that matter and energy are derivative phenomena—takes some getting used to.

Still, the idea has slowly gained credibility since Fredkin first articulated it decades ago. It is at the foundation of the new sciences of complexity that a bevy of Nobel laureates are eagerly exploring at the famous Santa Fe Institute. And if you tilt your head and squint at the idea from just the right angle, it displays a certain undeniable plausibility.

Here is how Fredkin described his theory in an interview with science writer Robert Wright:

> What I'm saying is that at the most basic level of complexity an information process runs what we think of as physics. At the much higher level of complexity life, DNA—you know, the biochemical functions—are controlled by a digital information system. Then, at another level, our thought processes are basically information processing.[6]

That seems eminently reasonable, indeed, uncontroversial. But then Fredkin goes off the deep end, plunging into treacherous and murky metaphysical waters that most scientists avoid like the plague. It's not that Fredkin is simply arguing that computation and computers are useful metaphors that can aid our understanding of nature (similar to the metaphor of artificial selection that helped Darwin formulate his theory of evolution through natural selection). No, Fredkin believes that the universe *really is* a giant mainframe. As Wright puts it:

> Fredkin believes that the universe is very literally a computer and that it is being used by someone, or something, to solve a problem. It sounds like a good news/bad news joke: the good news is that our lives have purpose; the bad news is that their purpose is to help some remote hacker estimate pi to nine jillion decimal places.[7]

What is the cosmic computer actually computing? What is it trying to figure out? Fredkin admits he has no idea and indeed insists that there's no way to discover the answer to this ultimate question—or, for that matter, the identity of the Cosmic Programmer—without waiting around for the program that constitutes the evolving universe to complete its entire operating cycle. As Fredkin puts it:

> There is no way to know what the future is any faster than running [the universe] to get to that [future]. Therefore, what I'm assuming is that there is a question and there is an answer, okay? I don't make any assumptions about who has the question, who wants the answer, anything.[8]

The longer Fredkin goes on explaining the implications of his theory, the more he seems to veer off into a weird techno-religious zone that bears an uncanny resemblance to the scientifically disreputable intellectual territory populated by the Intelligent Design crowd:

> Every astrophysical phenomenon that's going on is always assumed to be just an accident. To me, this is a fairly arrogant position, in that intelligence—and computation, which includes intelligence, in my view—is a much more universal thing than people think. It's hard for me to believe that everything out there is just an accident.... What I can say is that it seems to me that this particular universe we have is a consequence of something I would call intelligent.... There's something out there that wants to get the answer to a question.... Something that set up the universe to see what would happen.[9]

Especially quirky is Fredkin's fervid insistence that the religious concept of *soul* can be reformulated in terms of computational theory:

> The *soul* in every living thing is the informational part of that thing that is purposefully engaged in the informational aspects of its ability to be conceived or germinate, grow with cells differentiating, grow further in size, move, make use of sensory information, react reflexively, learn, behave instinctually, think intelligently, communicate with other beings, teach, reproduce, evolve and in general carry out informational interactions starting with the combining of parental DNA, informational interactions with itself, with things external to itself through senses, actions, constructions, creations and communications, and with its progeny through contributed DNA. A *soul* can learn from experience, from reflection or by being taught by other *souls*. In short, a soul can teach other *souls*.[10]

If this sounds like the ranting of some weirdo who has cobbled together an intellectual hodgepodge consisting of bits and pieces of the spiritual teachings of the Dalai Lama, the sermons of Pat Roberts, and lessons learned in Computer Science 101, then you need to remind yourself that truly original thinkers often stray into what professional skeptic Michael Shermer calls the borderlands of science—or even far beyond the legitimate borders of science into the realm of spiritualism and rampant religious speculation.

Isaac Newton's theological ruminations, for example, dwarf in page count the content of his famous *Principia Mathematica*.

The Secret Religious Writings of Isaac Newton

Isaac Newton is the father of modern physics, and arguably the greatest scientific genius who ever lived. His storied mathematical depiction of the force of gravity in the *Principia Mathematica* required him to first invent calculus. The elegance and simplicity of his formulations furnish the standard by which the quality, and indeed the very beauty of scientific work, has been judged ever since.

It comes, then, as a bit of a shock to learn that Newton's deepest passion was not unraveling the mysteries of the ubiquitous gravitational force, but rather attempting to discover hidden messages in the Bible. The secret writings in which he distilled the fruits of his biblical scholarship are only now coming to light.

For instance, Newton wrote a 300,000-word commentary on the Book of Revelation (available online at *www.newton project.ic.ac.uk*). Somewhat ominously, near the end of his life he casually predicted that the most dramatic events of the Apocalypse would begin to take place in the year A.D. 2060.

More recently, respected cosmologist Frank Tipler published a book entitled *The Physics of Immortality*[11] that seeks to prove the teachings of the Christian faith on the basis of mathematical physics. And Francis Crick, the Nobel laureate who co-discovered the molecular structure of DNA—arguably the most important triumph in biology since the publication of Charles Darwin's *The Origin of Species*—enthusiastically endorsed the oddball notion that intelligent extraterrestrials deliberately seeded Earth with primitive life (the idea is called directed panspermia). So perhaps genius should be accorded the privilege of occasional eccentricity.

Make no mistake, Fredkin is a genuine genius. As MIT computer scientist Marvin Minsky put it, the creators of new scientific paradigms are unlike most working scientists. For such radical innovators, "there's no point talking to anyone but a Feynman or an Einstein or a Pauli. The rest [of working scientists] are just Republicans and Democrats."[12]

Seth Lloyd and the Cosmos as a Quantum Computer

The newest and most sophisticated advocate of the view that the entire cosmos should be regarded as a massive computer is Seth Lloyd. Lloyd is a professor of mechanical engineering at MIT, principal investigator at the Research Laboratory of Electronics, and designer of the world's first feasible quantum computer. He frequently contributes to prestigious journals such as *New Scientist* and *Scientific American.*

As with Wolfram and Fredkin, Lloyd thinks the cosmos is a giant computational device. But he takes this proposition one step further, asserting that the universe is a huge *quantum computer.*

Quantum Computation

The difference between a quantum computer and an ordinary PC is that the former exploits all the weird aspects of quantum physics to conduct super-fast calculations that are simply impossible on a traditional digital computer. For instance, rather than churning through various numerical combinations that might crack a top-secret National Security Agency code, a quantum computer can explore all the possibilities simultaneously. (It is hardly surprising that the NSA is one of the top sources of funding for quantum computing research.)

Although promising, quantum computing faces daunting engineering challenges, paramount among them the difficulty of keeping the computing elements of the device (called *qubits*) in a delicate state of *quantum entanglement* until the computational task at issue has been completed.

As Lloyd puts it in his 2006 book *Programming the Universe,* "The history of the universe is...a huge and ongoing quantum computation. The universe is a quantum computer."[13]

But what precisely is the universe computing? Lloyd's answer is straightforward, if a trifle cryptic:

> [The universe] computes itself.... As soon as the universe began, it began computing. At first the patterns it produced were simple, comprising elementary particles and establishing the fundamental laws of physics. In time, as it processed more...information, the universe spun out...more...complex patterns, including galaxies, stars, and planets. Life, language, human beings, society, culture—all owe their existence to the intrinsic ability of matter and energy to process information.[14]

Indeed, Lloyd believes that his insight explains one of the great mysteries of nature: Why is the cosmos so complex? Why didn't the universe just remain a uniform, dilute soup of featureless matter and energy after the Big Bang? Where did all the evident diversity come from—matter clumped into stars; stars grouped into giant galaxies; and on at least one planet orbiting an ordinary sun, matter organized into the kind of dynamic, self-reproducing order that is the signature of life? What is the source of this puzzling variety, order, and organization, particularly the overwhelming and self-renewing complexity of living things?

Here is Lloyd's response:

> The computational capability of the universe explains one of the great mysteries of nature: how complex systems such as living creatures can arise from...simple physical laws.... The digital revolution under way today is...the latest in a...line of information-processing revolutions stretching back...to the beginning of the universe itself. Each revolution has laid the groundwork for the next, and all information-processing revolutions since the Big Bang stem from the intrinsic information-processing ability of the universe. The computational universe *necessarily* generates complexity. Life, sex, the brain, and human civilization did not come about by mere accident.[15]

Quantum Evolution

The relationship between quantum physics, life, and evolution has fascinated scientists and philosophers ever since the dawn of the quantum revolution in the early 20th century. At first glance it would seem that these topics have little to do with one another—the physics of the micro-world and the phenomena of interest to evolutionary biologists would appear superficially to occupy widely separated niches on what quantum physicist Richard Feynman called the hierarchy of complexity that begins with the fundamental laws of physics and culminates in complex phenomena such as biology and culture. But a moment's reflection will show that this is not the case.

As Erwin Schrödinger, one the pioneers of quantum physics, pointed out in his seminal book, *What Is Life?*, atomic activity and structure at the micro-level is responsible for supervising biological activity at the macro-level:

> [I]ncredibly small groups of atoms, much too small to display exact statistical laws, do play a dominating role in the very orderly and lawful events within a living organism. They have control of the observable large-scale features which the organism acquires in the course of its development, they determine important characteristics of its functioning; and in all this very sharp and very strict biological laws are displayed.[16]

As Schrödinger notes, the key factor that accounts for the sharpness and strictness of biological laws—and in particular, for the phenomenon of inheritance through incredibly accurate replication of genetic material across multiple generations—is the discreteness of physical reality at its smallest scales, as revealed by quantum physics:

> In the light of present knowledge, the mechanism of heredity is closely related to, nay, founded on, the very basis of quantum theory.[17]

What is crucial for biology is the fact that "a number of atomic nuclei, including their bodyguards of electrons, when they find themselves close to each other, forming 'a system,' are unable by their very nature to adopt any arbitrary configuration we might think of. Their very nature leaves them only a very numerous but discrete number of 'states' to choose from."[18] This quantization of reality at the atomic level is the factor that enables DNA to function, with nearly flawless accuracy through millions of replications, as a superb digital memory chip. If reality were not quantized and discrete at the molecular level, DNA simply could not function as it does; it could not serve as life's reliable shepherd across immense stretches of geological time. Absent this capacity on the part of DNA, there could be no macro-phenomenon of evolution (at least as we know it) and perhaps no such thing as life—certainly not complex life. And so the quantum nature of reality at the molecular level must be added to the list of qualities that render our universe so mysteriously bio-friendly!

But the linkage between quantum physics and biological evolution may go deeper still. In a pioneering book entitled *Quantum Evolution*, molecular geneticist Johnjoe McFadden suggests that quantum mechanical rules provide a possible way of overcoming "the huge improbability of the first self-replicator"[19] and thus a possible explanation for the origin of life. His theory, which remains highly speculative, is essentially that the key molecules that make up living creatures were in a quantum superposition before they decohered into a linked, autocatalytic system that became the first living entity.

Life, mind, and human civilization are no mere accidents? Once again, this is starting to sound suspiciously like Intelligence Design rhetoric. But let's defer judgment pending a closer look.

As noted previously, the key feature distinguishing Seth Lloyd's vision of the basic nature of nature from those of Wolfram and Fredkin, is that Lloyd believes the cosmos is not a classical digital computer, but a giant *quantum* computer, capable of exploiting all the weirdness of quantum mechanics.

What kind of odd tricks does the cosmic computer have access to? All the strange quantum phenomena that sub-atomic particles exhibit—the ability of these tiny things to be in two places at once and to behave simultaneously as waves and particles, the odd sensitivity of quantum phenomena to the very act of observation (as if observers somehow influence the behavior of electrons and other inanimate physical entities just by staring at them) and, perhaps most mysterious of all, the phenomenon of "spooky action at distance" (Einstein's famous phrase) by which the act of observation of one entangled particle instantaneously influences the physical state of a second entangled particle, even though the two entangled particles are light-years apart!

These are the same oddities of quantum physics that Lloyd and other computer scientists are attempting to harness in designing artificial quantum computers that would run rings around conventional digital processors. The crucial advantage of quantum computers is that whereas the processor inside the PC that sits on your desk must process bits sequentially, a quantum computer exploits the peculiar capacity of a quantum bit—a *qubit*—to register two states (such as the zero and one of binary arithmetic) simultaneously. This means, as Lloyd notes, that "a quantum computer can perform millions of computations simultaneously."[20] As a consequence, a quantum computer can, in principle, quickly perform tasks that would require a virtual eternity on the part of even the fastest conventional supercomputer.

This is the key to Seth Lloyd's crucial insight: Only a universe that functions as a quantum computer is capable of generating all the marvelous order, diversity, and complexity that we observe around us. As Lloyd points out, "quantum mechanics, unlike classical mechanics, can create information out of nothing."[21] (Now that's what I call spooky!) The result is that, as the universe engages in quantum computation over billions of years, it continually creates more opportunities for its constantly diversifying elements and structures to interact with one another in ever more novel and complicated ways.

In effect, the cosmos teaches itself how to grow ever more complex as it ages. Nobody is telling the universe how to engage in this ultimate feat of self-enlightenment. The universe simply creates its own instructional syllabus *ex nihilo* as it ticks through the sequence of microscopic quantum events comprising its multi-billion-year history. Almost miraculously, there is embedded in the quantum-processing capability of the cosmos an eerie proclivity for generating what complexity theorist Stuart Kauffman calls "order for free."[22] As Lloyd puts it:

> The computational universe paradigm allows new insights into the way the universe works. Perhaps the most important new insight afforded by thinking of the world in terms of information is the resolution of the problem of complexity. The conventional mechanistic paradigm gives no simple answer to the question of why the universe...and life on Earth...is so complex. In the computational universe...the innate information-processing power of the universe systematically gives rise to all possible types of order...[23]

Perhaps the most startling claim made by Seth Lloyd is that his quantum computational cosmic paradigm may offer a novel way of approaching what is perhaps the toughest challenge in contemporary physics: figuring out how to unify the seemingly incommensurable realms of quantum mechanics—which flawlessly predicts the behavior of nature at the very smallest scale (at the level of interactions of atoms and subatomic particles)—and general relativity, which accurately explains the behavior of natural objects at nature's largest scale. The attempt to merge these two explanatory regimes—to formulate a satisfactory theory of *quantum gravity*—has resulted in a deluge of complicated and ingenious exercises in mathematical speculation, including string theory (and its successor, M-theory), as well as a competing approach called loop quantum gravity.

But Lloyd has made an audacious suggestion: Perhaps the string and loop theorists have been, in effect, gazing through the wrong end of the telescope. In a 2005 scientific paper titled "A theory of quantum gravity based on quantum computation"[24] he proposed that "the geometry of space-time is a construct, derived from the underlying quantum information processing."

The beauty of this approach is that it slyly evades the necessity of directly reconciling quantum physics and relativity. As Lloyd puts it:

> Unlike conventional approaches to quantum gravity such as string theory...[and] loop quantum gravity...the theory proposed here does not set out to quantize gravity directly. What is quantized here is information: all observable aspects of the universe, including the metric structure of space-time and the behavior of quantum fields, are derived from and arise out of an underlying quantum computation. The form that quantum fluctuations in geometry take can be calculated directly from the quantum computation. To paraphrase Wheeler, "it from qubit."[25]

John Wheeler's Radical Idea: It From Bit

No scientist has formulated the primacy of information over matter more boldly than Princeton physicist John Wheeler. As he put it in an essay titled "It From Bit":

> *It from bit*. Otherwise put, every it—every particle, every field of force, even the space-time continuum itself—derives its function, its meaning, its very existence entirely—even if in some contexts indirectly—from the apparatus-elicited answers to yes or no questions, binary choices, *bits*. It from bit symbolizes the idea that every item of the physical world has at bottom—at a very deep bottom, in most instances—an immaterial source and explanation; that what we call reality arises in the last analysis from the posing of yes-no questions, and the registering of equipment-evoked responses; in short, that all things physical are information-theoretic in origin and this is a *participatory universe*.[26]

The essence of Wheeler's vision—a radical extrapolation from the Copenhagen interpretation of quantum physics—is that the very act of observation is necessary to summon the universe into existence and that it literally does not exist until it is observed. Few scientists adhere to this extreme position.

Hitching a Ride on the Cosmic Software

An interesting but infrequently discussed aspect of the universe-as-computation paradigm promoted by Seth Lloyd, Edward Fredkin, and Stephen Wolfram is that the lives and thoughts of human beings are necessarily part and parcel of the cosmic program. So, too, are the computational operations of mechanical artifacts (such as computers) that humans design and build. Indeed, Lloyd asserts that when we build and operate computers (quantum or traditional), what we are really doing is "hacking existing systems. You're hijacking the computation that's already happening in the universe, just like a hacker takes over someone else's computer."[27]

As argued at length in my first book, *Biocosm*, and as discussed in Chapter 8 of this book, the deepening involvement of human and transhuman intelligence in the ongoing quantum computation that Seth Lloyd contends is the foundational reality of our universe has profound implications for the future state of the cosmos and for its ultimate destiny. But for now I want to focus on the intermediate stages in the unfolding saga of cosmic quantum computation. Specifically, I want to investigate whether there may be strategies by which we can more efficiently hijack the computation that's already happening in our universe and harness it for our own selfish purposes.

Can we become clever cosmic hackers? Can we learn how to stick out our thumbs and hitch a ride on the Software of Everything, thereby exploiting it directly for our own ends? To do so, will we need to discover or compose a nonfiction version of the comic classic *The Hitchhiker's Guide to the Galaxy*?[28] Or might the process of clambering aboard the great cosmic software express possibly be as simple and straightforward as mindlessly emulating Darwinian natural selection? Could the most efficient means of hijacking the ongoing cosmic computation be to somehow persuade that computation to highjack itself for our benefit?

Intelligent Design versus the Evolutionary Model of Software Creation

We are accustomed to thinking of computer software as a meticulously fabricated artifact, constructed by hordes of brainiacs at Microsoft or IBM or Apple. Millions of lines of code are engineered into useful products by careful programmers who attempt to anticipate how stupendously complex ensembles of software code will interact. But this paradigm of software development—call it the Intelligent Design model—is just one way in which software can be created. This approach has produced wonderfully useful products that have transformed society and dramatically increased economic productivity around the globe, though it is slowly giving ground to a radically different method of software creation.

A key factor motivating this transformation is a kind of limiting principle attributed to Nicklaus Wirth that is the flip side of Moore's law. (Moore's law asserts that approximately every 18 months, computers double in the speed with which they process information.[29]) Wirth's Law was described (as follows) in a recent essay published in *Skeptic* magazine:

> There's another "law," this one attributed to Nicklaus Wirth: software gets slower faster than hardware gets faster. Even though, according to Moore's Law, your *personal computer* should be about a hundred thousand times more powerful than it was 25 years ago, your *word processor* isn't. *Moore's Law doesn't apply to software.*[30]

The problem, in a nutshell, is that human beings still play the central role in traditional software fabrication. And the human mind, however well-educated and exceptionally brilliant it might be, still bears the limitations on speed and capacity that are the ineradicable legacy of our evolutionary past. In Darwin's wonderful phrase, "We must...acknowledge...that man...with his god-like intellect which has penetrated into the movements and constitution of the solar system—with all these exalted powers—man still bears in his bodily frame the indelible stamp of his lowly origin."[31] As with the human body, so too with the human brain: It is inescapably a biologically evolved organ incapable of transcending the inherent limitations of its evolutionary past.

But what if the human beings could be taken out of the loop in the software design process? What if software construction could somehow be automated? Might the limitations of Wirth's Law then be circumvented? And could the raw power of natural selection be harnessed directly in a nonbiological setting? This is the exciting possibility that a fringe group of software engineers are currently exploring.

Their radically novel approach to software design sports a variety of futuristic labels, such as evolutionary computation, genetic programming, and genetic algorithm development. It seeks to directly harness the awesome power of Darwinian natural selection to literally *evolve* software systems—incredibly complex programs that often defy the ability of their human stewards to understand exactly how or why they work. As genetic programming expert Lee Spector wrote recently with regard to the development of a prize-winning genetic program whose emergence he oversaw:

> My entry involved the evolution of quantum computing circuits, which are difficult for humans to understand or design. More to the point, they are extremely difficult for *me* to understand or design, and I could never have produced the results on my own. I am not a designer equal to that task, but evolution is. I created the "primordial ooze" out of which quantum circuits could grow, and I wrote the programs for random variation and selection. But evolution did the heavy lifting.[32]

The basic approach of genetic programmers is to set a high-level goal—creation of a new electronic circuit, for instance, or a new kind of antenna for NASA space missions—and then set loose a swarm of software agents to attack the problem. Those strings of software that show some initial success are allowed to breed with other successful software agents and propagate software progeny. The less-adept software agents are mercilessly culled from the computer's memory, with the human overseer playing the role of grim reaper.

Progeny of the successful software agents then attack the same design challenge. The less-fit agents in the second software generation—fitness always being measured in terms of relative ability to solve the target challenge (for instance, design of a new electronic circuit with specified performance characteristics)—are weeded out and, once more, the more successful agents are allowed to engage in a kind of simulated mating ritual to reproduce.

This Darwinian process is repeated thousands of times until an arbitrary decision is made by the human experimenter that a solution developed by the final generation of evolved software agents is adequate to the task at hand. Then the experimenter turns off the computer and the mini-universe of frantically competing computer programs simply winks out in quiet oblivion, ending not with a bang or a whimper, but an empty computer screen.

This approach is a way of directly harnessing the stupendous power of the ongoing cosmic computation. The "heavy lifting" (in Spector's felicitous phrase) is performed not by human software engineers but by interacting and competing strings of information. Humans may set the rules of the game and specify the fitness criteria that determine which software entities will survive to reproduce and which will perish, but no human mind designs the winning solution. No human intelligence guides the specific pathway taken by the Darwinian process.

Genetic programming possesses some inherent advantages over the traditional Intelligent Design approach to software fabrication.

Software companies using genetic programming techniques may not need to pay as many expensive salaries to highly educated software engineers as companies using the traditional design approach. The ongoing trend toward the use of voice recognition software in lieu of call centers staffed by human beings is a vivid demonstration that it can be far more cost-effective to employ a machine rather than a person to perform complicated informational tasks once regarded as beyond the capacity of artificial systems.

Unlike traditional software programming, genetic programming will, in the future, increase its capabilities exponentially in tandem with the ongoing unfolding of Moore's law, whereas human software engineers will be constrained by limits on human intelligence and ingenuity imposed by our evolutionary past.[33] As one leading expert in the field put it:

> Genetic programming is able to take advantage of the exponentially increasing computational power made available by iteration of Moore's law. This...suggests that genetic programming may deliver increasingly more significant results in the future.[34]

In short, genetic programming *does* obey Moore's law. The limits of Wirth's Law, which essentially reflect the inherent limitations of the human mind, do not apply.

Perhaps most intriguing, genetic programming is unconstrained by the habits of thought and subtle intellectual biases of human programmers. Because they mindlessly explore the entire space of possible solutions to a particular problem, the evolving software entities can come up with solutions that are, well, just plain alien. For instance, NASA used genetic programming to come up with an optimal design for a girder to be used on the International Space Station. As reported in *U.S. News and World Report*, the result was straight out of a science fiction novel:

> There emerged, from 15 generations and 4,500 different designs, a truss no human engineer would design. The lumpy, knob-ended assembly [resembled]...a leg bone, irregular and somehow organic. Tests on models confirm its superiority to human-designed ones as a stable support. No intelligence made the designs. They just evolved.[35]

Another startling example of the sheer strangeness of successful genetic programming is computer code that was evolved to help a patient control a prosthetic hand on the basis of erratic nerve signals picked up by electrodes taped to the patient's wrist. The evolved software somehow mysteriously analyzed the nerve signals and translated them, with perfect accuracy, into the hand movements that the patient wanted to make.

But here's the really weird part—the aspect of the experiment that should make the hairs on the back of your neck tingle. The method by which the evolved software performed this amazing feat was utterly beyond the comprehension of the human researchers. As reported in *Scientific American*:

> The evolved code [was] as messy and inscrutable as a squashed bug. [The] gesture-predicting program consists of a single line so long that it fills an entire page and contains hundreds of nested parenthetical expressions. It reveals nothing about why the thumb moves a certain way—only that it does.[36]

From Lottery Tickets to Artificial Darwinism: The Amazing Journey of John Koza

John Koza is a burly guy who somewhat resembles the middle-aged truck driver you might run into at a local diner. But looks can be deceiving. Koza, a consulting professor at Stanford with positions in both the School of Medicine and the School of Engineering, is the world's leading expert in the field of genetic programming. He invented the field, and literally wrote the book on the subject.[37] He is so committed to achieving progress in this type of computer programming that he has installed a massively parallel computer in a spare room of his hilltop home in Silicon Valley, so that he can keep his research going constantly.

Koza previously served as cofounder and chief executive officer of Scientific Games, Inc., where he invented the instant scratch-off lottery ticket. For this achievement he has been inducted into the Lottery Industry Hall of Fame, established by the Public Gaming Research Institute.

I met Koza in the fall of 2003 when we both spoke at the Accelerating Change Conference at Stanford. The conference was a non-stop intellectual feast for people who like to think about what the future holds for the human race. Speakers included computer scientist Ray Kurzweil, nanotech expert and venture capitalist Steve Jurvetson, and popular science author Robert Wright.

John Koza was a clear stand out in this crowd of futurists and visionaries. He began with a somewhat daunting definition of a "genetic algorithm," which can be thought of as the workhorse of genetic programming:

> The *genetic algorithm* is a probabilistic search algorithm that iteratively transforms a set (called a *population*) of mathematical

> objects (typically fixed-length binary character strings), each with an associated fitness value, into a new population of offspring objects using the Darwinian principle of natural selection and using operations that are patterned after naturally occurring genetic operations, such as crossover (sexual recombination) and mutation.[38]

As Koza warmed to his topic, the sheer revolutionary magnitude of what he was saying slowly began to sink in. This was not just a new way of programming computers. The dream Koza was chasing was the Holy Grail of computer science, expressed presciently by Arthur Samuel—founder of the field of machine learning—in 1959: "How can computers be made to do what needs to be done, without being told exactly how to do it?"[39]

As Koza discussed the potential of genetic programming, it became clear that this former gambling industry pioneer was not focused on abstract proof-of-concept laboratory results (what researchers call *toy problems*), but real-world software solutions that compare favorably with *the best outcomes that human designers can achieve!* Koza had his sights set on the target of commercial success, where the stakes are not bragging rights at the Stanford faculty club, but potentially billions of dollars in profits from artificially evolved software programs, electrical circuit designs, and novel pharmaceutical products.

That's why Koza describes his demanding criterion for measuring success in the field of genetic programming as a requirement that artificially evolved programs yield human-competitive results. In other words, to be considered successful on Koza's exacting terms, evolved software must perform at least as well as any program that a human software engineer could design.

Remarkably, there are now a number of instances where genetic programming has, in his words, "automatically produced a computer program that is competitive with human performance."[40] These include "reinventions" by genetic programs of previously patented human inventions and the design, from scratch, of at least two entirely new patentable inventions. The implications are truly mind-boggling. As Koza puts it:

> The fact that genetic programming can evolve entities that are competitive with human-produced results suggests that genetic programming may possibly be used as an "invention machine" to create new and useful patentable inventions. In this connection, evolutionary methods, such as genetic programming, have the advantage of not being encumbered by preconceptions that limit human problem-solving to well-traveled paths.[41]

The Coming Fusion of Genetic Programming and Quantum Computation

Although John Koza has set his sights on routinely achieving "human competitive" results through genetic programming, it is hard to escape the uneasy sense that this is not the end of the story. Once this threshold has been achieved, why shouldn't artificial Darwinian evolution thereafter exceed the limits that biological evolution has placed on the capabilities of even the most talented human computer programmer? In short, why isn't it reasonable to anticipate that Koza's genetic programming marvels will eventually exceed the capacity of the human mind?

In fact, many prominent futurists believe that the moment when self-evolved software will acquire transhuman mental power—an epochal event that computer scientist Ray Kurzweil calls a computational singularity—is near. But what has received less attention is the potential contribution to this development of quantum computational techniques such as those discussed by Seth Lloyd, as well as the inherent synergy of genetic programming and quantum computation.

Moreover, this potent fusion—the use of Darwinian natural selection processes to evolve ever more powerful software algorithms employing a quantum computer as a platform—would seem to tap into the same kind of "quantum magic" that British physicist Roger Penrose believes is the key to the ability of tiny structures in the human brain, called microtubules, to generate the elusive phenomenon of human consciousness.

Microtubules, Quantum Computing, and Consciousness: The Purported Link

In *The Emperor's New Mind*[42] Roger Penrose, the Emeritus Rouse Ball professor of mathematics at the University of Oxford, took on the promoters of strong artificial intelligence by arguing that there was something unique about human consciousness that could never be duplicated in a computer:

> According to [the] perception [of believers in the possibility of strong artificial intelligence], *all* aspects of mentality (including conscious awareness) are merely features of the *computational* activity of the brain; consequently, electronic computers should also be capable of consciousness, and would conjure up this quality as soon as they acquire sufficient computational power and are programmed in an appropriate way. I do my best to express, in a dispassionate way, my scientific reasons for disbelieving this perception,

> arguing that the conscious aspects of our minds are *not* explicable in computational terms and moreover that conscious minds can find no home within our present-day scientific world-view.[43]

In a dazzling display of multidisciplinary erudition, Penrose draws on quantum theory, Gödel's incompleteness theorem, cosmology, relativity theory, Platonic philosophy, and a host of other sources to buttress his controversial view that the human mind is something more—much more—than a mere "meat computer." The essence of his argument is that human consciousness must, as a matter of logical necessity, arise from some peculiar class of quantum phenomena operating at the border between "the submicroscopic world of quantum physics [and] the macro-world of classical physics"[44] and that there were special phenomena to which a mere computer could never gain access.

In a subsequent book (*Shadows of the Mind*)[45] Penrose seeks to remedy a perceived weakness of *The Emperor's New Mind* by attempting to actually locate the physical situs in the human brain responsible for generating consciousness:

> One of the major shortcomings of [*The Emperor's New Mind*] is, perhaps, that when I wrote it I knew of no place in the brain where it could be plausibly argued that the "large-scale quantum coherence" could take place that would be needed for the application of the ideas [set forth in the book].... One of [the] scientists [who read the book] was Stuart Hameroff, who acquainted me with the cell's cytoskeleton and its microtubules—structures of which I had been deplorably ignorant previously! He also informed me of his own ingenious ideas concerning a possible role for microtubules, within the brain's neurons, in relation to the phenomenon of *consciousness*. It seemed to me that the most plausible place for the kind of large-scale quantum coherent action that my argument required was indeed within microtubules.[46]

Despite Penrose's undeniable scientific and mathematical brilliance, his views about the nature and source of human consciousness have gained few adherents.

If Penrose is correct, could quantum computation be the instrumentality that will eventually endow highly evolved software entities "living" in quantum computers with something resembling human consciousness? Even if he is incorrect (and he is almost certainly wrong, at least in the view of many leading neuroscientists and quantum physicists[47]), the sheer computational muscle of quantum processors will inevitably speed up, by orders of magnitude, the process of artificial Darwinian evolution that is the basic engine of genetic programming.

One scientist who has begun to think deeply about the linkage between genetic programming and quantum computation is Lee Spector, a professor of computer science at Hampshire College and author of *Automatic Quantum Computer Programming: A Genetic Programming Approach.*[48] Spector's book focuses on the use of genetic programming to aid in designing circuits for quantum computers. This is a vitally important topic, because computer scientists lack a thorough understanding of how to optimally harness the awesome power of quantum computation for useful purposes. Quantum programs and quantum algorithms are fiendishly difficult to understand and even harder to write down. For this reason, only a handful of quantum computational algorithms have been formally documented. As three computer scientists noted in a 2004 paper:

> Despite the scientific and technological importance of quantum computation, few quantum algorithms faster than the classical ones have been discovered.... This is due to the fact that the generation of such algorithms or circuits is difficult for a human researcher. They are unintuitive, mainly due to quantum mechanics features like [quantum] entanglement and [quantum] collapse.[49]

The daunting challenge of figuring out how to best program a quantum computer presents a perfect opportunity to bring to bear the special attributes of genetic programming, which has already yielded several important new quantum computational algorithms. The key advantage of using genetic programming to develop algorithms for quantum computers is that a genetic program will, true to Darwinian principles of random search, blindly search for all possible programming solutions, including nonobvious approaches to algorithm design that would never have occurred to a human programmer.

Whereas the use of genetic programming to design quantum computing algorithms is well underway, a project still in its infancy is exploration of the use of quantum computers to *run* genetic programs, including genetic programs that would search for new and improved algorithms with which to run quantum computers. Work has barely begun on this exciting topic, but it is already clear that, at the very least, a genetic program running on a quantum computer will benefit from orders-of-magnitude acceleration in processing speed. However, as Spector has remarked:

> One would like to get more than this, and I believe there are reasons to believe that one can. These includes tantalizing similarities between features of quantum computing models and features of evolutionary search models, ranging from the

> uses of randomness to the ubiquity of exponentially branching search spaces that can be searched in parallel.[50]

The potential of marrying up these two cutting-edge computing technologies has also been noted by three leading Brazilian researchers who believe that, despite the fact that little work has been done to date in the emerging field of quantum evolutionary programming, it is an exceptionally exciting and potentially very fruitful subject of research.[51]

Could this nascent technique be the method by which human beings and their clever mechanical artifacts (that is to say, advanced quantum computers) will ultimately be able to fully hijack the ongoing quantum computation that powers the physical evolution of the universe itself? And by employing this technique, will we stumble onto the god-like power to create artificial life and genuine artificial intelligence? And if it should emerge from our experiments, either deliberately or accidentally, will artificial intelligent life forms, sooner or later, overshadow their mortal creators?

Samuel Butler's Prescient Prophecy of Artificial Life

It is interesting to note that the most eloquent prophecy on these weighty topics remains the very first, delivered by Samuel Butler, just four short years following the publication of Charles Darwin's *The Origin of Species*:

> There are few things of which the present generation is more justly proud than the wonderful improvements which are daily taking place in all sorts of mechanical appliances.... But what would happen if technology continued to evolve so much more rapidly than the animal and vegetable kingdoms? Would it displace us in the supremacy of earth? Just as the vegetable kingdom was slowly developed from the mineral, and as in like manner the animal supervened upon the vegetable, so now in these last few ages an entirely new kingdom has sprung up, of which we as yet have only seen what will one day be considered the antediluvian prototypes of the race.... We are daily giving [machines] greater power and supplying by all sorts of ingenious contrivances that self-regulating, self-acting power which will be to them what intellect has been to the human race.[52]

Chapter 2

Stages on Artificial Life's Way

Who will be man's successor? To which the answer is:
We are ourselves creating our own successors.
Man will become to the machine what the horse and the dog are to man;
the conclusion being that machines are, or are becoming, animate.

—Samuel Butler, 1863 letter titled "Darwin Among the Machines"[1]

In the summer of 2000, I was invited to deliver a scientific paper at the Seventh International Artificial Life Conference. The conference took place on the beautiful campus of Reed College in Portland, Oregon. Amid towering Douglas fir trees and expanses of well-tended lawns, a motley crew of computer scientists, physicists, NASA officials, theoretical biologists, national weapons lab researchers, and assorted freethinkers assembled to reflect on one of nature's most perplexing puzzles: What is life?

The question has a famously long and fractious history, stimulating poets and skeptics, philosophers and quantum mechanics, biologists and mystics, to offer a bewildering potpourri of radically different explanations. Life is a unique phenomenon and fundamentally different from non-life, opined French philosopher Henri Bergson. Bergson theorized that life is pushed irresistibly to ever-higher levels of evolutionary accomplishment by a mysterious vital force (*élan vital*) that is entirely absent from non-living matter.

Nonsense, retorted skeptic Robert Morison. The term *life* is just a linguistic convention we employ in order to describe a special class of material objects that possess some unusual thermodynamic and behavioral characteristics. Beyond the fact that they share certain common properties, *living* things are indistinguishable from *lifeless* stones:

> Life is not a thing or a fluid any more than heat is. What we observe are some unusual sets of objects separated from the rest of the world by certain peculiar properties such as growth, reproduction, and special ways of handling energy. These objects we elect to call "living things."[2]

Life's secret weapon, concluded quantum physics pioneer Erwin Schrödinger in a pithy book entitled *What Is Life?*, is its unique ability to metabolize—to export disorder to the surrounding environment in the form of radiated heat and excrement while importing order from that environment in the form of food and energy:

> It is by avoiding the rapid decay into the inert state of "equilibrium" that an organism appears so enigmatic; so much so, that from the earliest times of human thought some special non-physical or supernatural force (*vis viva*, entelechy) was claimed to be operative in the organism, and in some quarters is still claimed. How does the living organism avoid decay? The obvious answer is: By eating, drinking, breathing and (in the case of plants) assimilating. The technical term is *metabolism*.... What then is that precious something contained in our food which keeps us from death? That is easily answered. Every process, event, happening—call it what you will; in a word, everything that is going on in Nature means an increase of the entropy of the part of the world where it is going on. Thus a living organism continually increases its entropy...thus tends to approach the dangerous state of maximum entropy, which is death. It can only keep aloof from it, i.e., alive, by continually drawing from its environment negative entropy.... What an organism feeds upon is negative entropy. Or, to put it less paradoxically, the essential thing in metabolism is that the organism succeeds in freeing itself from all the entropy it cannot help producing while alive.[3]

Schrödinger's book was the inspiration for an entire generation of scientists who created, essentially from scratch, the vast scientific enterprise now known as molecular biology. In particular, Schrödinger anticipated and inspired the epic achievement of James Watson and Francis Crick: the discovery of DNA's structure.

What Is DNA?

The first serious speculation about the nature of the potent chemical substance that governs both the inheritance of genetically prescribed traits and the development of individual organisms from fertilized egg to mature adulthood came from Erwin Schrödinger. In *What Is Life?* he suggested that a "hereditary code-script"[4] written into the molecular fabric of chromosomes was responsible for the twin miracles of genetic transmission and ontogeny (the maturation of an organism during its life cycle). James Watson, who shared a Nobel Prize for discovering the double-helix structure of DNA, has revealed that it was Schrödinger's conjecture that inspired his insight:

> Schrödinger argued that life could be thought of in terms of storing and passing on biological information. Chromosomes were thus simply information bearers.... Schrödinger's book was tremendously influential. Many of those who would become major players in Act I of molecular biology's great drama, including Francis Crick (a former physicist himself), had, like me, read *What Is Life?* and been impressed.[5]

Watson drew an important philosophical conclusion from his discovery of the molecular foundation of life and heredity—that life was not all that different from non-life, and that living matter harbors no inherently inscrutable secrets:

> Our discovery put an end to a debate as old as the human species: Does life have some magical, mystical essence, or is it, like any chemical reaction carried out in a science class, the product of normal physical and chemical processes? Is there something divine at the heart of a cell that brings it to life? The double helix answered that question with a definitive No.[6]

Ironically, his intellectual mentor (Schrödinger) reached precisely the opposite conclusion in *What Is Life?*, noting that life's defining characteristic—its capacity to produce and prolong the existence of an island of enduring order, surrounded and incessantly buffeted by a sea of randomness and entropy-driven disorder—is strong evidence of the existence of a "new type of physical law"[7] that governs the behavior of living matter.

Literary artists as well as scientists have weighed in on the perplexing issue of life's basic nature. Here is Thomas Mann's poetic invocation of this deep mystery in *The Magic Mountain*:

> What was life? No one knew. It was undoubtedly aware of itself, so soon as it was life; but it did not know what it was...it was not matter and it was not spirit, but something between the two, a phenomenon conveyed by matter, like the rainbow on the waterfall, and like the flame. Yet why not material?—it was sentient to the point of desire and disgust, the shamelessness of matter become sensible of itself, the incontinent form of being. It was a secret and ardent stirring in the frozen chastity of the universal; it was a stolen and voluptuous impurity of sucking and secreting; an exhalation of carbonic gas and material impurities of mysterious origin and composition.[8]

Modern approaches to the challenge of defining life are grouped into three basic categories: (1) explaining life as a collection of observed properties (such as the tremendous complexity of all living organisms and the key role played by natural selection in engineering the appearance of all organisms in the biosphere); (2) portraying life as metabolization (along the lines of Schrödinger's views discussed above); and (3) conceiving of life as evolution.[9]

The first two approaches have serious shortcomings. Any list-based definition of what is alive and what is not alive seems inherently arbitrary and potentially under-inclusive. How can we predict with any degree of confidence that life in general will always and inevitably exhibit the aggregate of traits we observe in the contemporary terrestrial biosphere? Is it possible that life was radically different in the distant past? Will it stubbornly adhere to its current manifestation in the distant future, or will it evolve into unrecognizably alien forms a million or a billion years from now? And will living forms exhibit the same basic characteristics as earthly critters if life is discovered in extraterrestrial settings?

The basic problem is that we are attempting to define life's basic characteristics on the basis of a single sample. What we desperately need is a *general* biology—a broad explanatory paradigm that gets at life's essence, wherever and whenever life may exist—rather than a simple catalogue of the common characteristics shared by terrestrial organisms.

The life-as-metabolism viewpoint also has a well-known weakness. As Mark Bedau, a leading artificial life researcher, put it:

> One drawback of metabolization as an all-encompassing conception of life is that many metabolizing entities seem not to be alive and not to involve life in any way. Standard examples include a candle flame, a vortex, and a convection cell.[10]

The third approach—the intriguing idea that life is not only a *product* of the pre-biotic evolution of increasingly complicated networks of organic macromolecules (a straightforward notion shared by mainstream biological theorists such as the Santa Fe Institute's colorful complexologist Stuart Kauffman[11])—but, in some mysterious sense, *is one and the same as the evolutionary process itself* is potentially revolutionary. Even more important, it may be right. And if it's right, it will shed new light on some of the deepest mysteries in biological science (including the puzzle of life's origin, which Charles Darwin famously predicted to be forever beyond the reach of the human mind).

The basic nature of life itself was the heady topic that the eclectic crowd of artificial life conventioneers (*a-lifers*, for short) who gathered at Reed College in the summer of 2000 to ponder and discuss. Their ambitious goal was to begin to construct a truly general biology—in the words of artificial life pioneer Christopher Langton, to understand not simply life as we know it, but *life as we do not know it*, and thus *life as it could be* in addition to life as it is here on Earth. Or, as Mark Bedau put it, "Artificial life seeks to understand all forms of life that could exist anywhere in the universe."[12]

The field of artificial life research is quite new—as an organized scientific discipline it dates back to a seminal conference on "Evolution, Games, and Learning: Models for Adaptation in Machines and Nature" held at Los Alamos National Laboratory in 1985[13]—but it has an ancient conceptual genealogy. The philosopher Thomas Hobbes offered an uncanny preview of the science of artificial life in his masterpiece *Leviathan* published in 1651:

> Nature (the Art whereby God hath made and governs the World) is by the *Art* of man, as in many other things, so in this also imitated, that it can make an *Artificial Animal*. For seeing life is but a motion of Limbs, the beginning whereof is in the principal part within; why may we not say that all *Automata* (Engines that move themselves by springs and wheels as doth a watch) have an artificial life?[14]

Contemporary artificial life researchers follow three basic paths in their attempt to construct a general biology and thereby understand life as it theoretically could be. As Bedau has put it:

> Contemporary artificial life (also known as "a-life") is an interdisciplinary study of life and lifelike processes. Its two most important qualities are that it focuses on the essential rather than the contingent features of living systems and that it employs synthetic methodologies. Artificial life's synthetic methodologies consist of studying life by synthesizing life-like behavior in artificial systems. Three different synthetic methods are being explored. "Soft" artificial life creates computer simulations or other purely digital constructions that exhibit life-like behavior, "hard" artificial life produces hardware implementations of life-like systems, and "wet" artificial life involves the creation of life-like systems out of biochemical substances. These synthetic methodologies help

> artificial life focus on essential rather than contingent properties of living systems, because they provide wide scope for exploring the nether boundaries of the landscape in which possible life forms can be found.[15]

Soft Artificial Life

Soft artificial life research is the best known of these three variations. It essentially consists of creating a special computer environment and then populating it with bits of software that compete with one another for various kinds of resources (such as memory space on a hard drive) and either thrive or perish depending on their relative fitness, defined as their ability to rapidly reproduce.

This is an approach similar to the genetic programming method previously described, except that the objective of a soft a-life experimenter is generally more open-ended than that of the designer of a genetic programming process. The goal of soft a-life research is not to breed software that will be adept at accomplishing a particular task designated in advance by the human researcher, but rather to turn the competing software creatures loose in a virtual computer ecosystem and then wait around to see what happens.

Software Life in the Wild: Could Computer Viruses Spontaneously Evolve in the Internet Ecosystem?

One of the limitations of software-based artificial life simulations is that none of them approach the level of dynamic evolutionary innovation that can be observed in every nook and cranny of the natural biosphere. Artificial software organisms are similar to farm-bred salmon; they lack the energy and verve of their wild cousins. As Mark Bedau writes, "Not a single artificial evolutionary model has unambiguously shown the sort of continually growing supple adaptation evident in the biosphere.... The problem seems to be that no existing model creates a continually unfolding accessible space of new kinds of adaptive innovations."[16]

Is the problem that the artificially synthesized computer ecosystems used to *simulate* the biosphere are too tame, too uncomplicated, and thus insufficiently challenging to generate species of lusty, vigorously evolving artificial life that resemble the natural carbon-based variety? Are the tame artificial creatures whose emergence we have witnessed to date the predictable spawn of the software counterpart of a fish farm or a cattle feedlot? Conversely, if artificial life forms could somehow break free to breed and evolve in a wild environment, would they become more genuinely lifelike or even come to life?

Such a wild environment exists. It is called the Internet. And some a-life visionaries are beginning to suspect that it will be in the dark heart of this untamed software jungle that truly living software organisms will evolve, perhaps as the distant progeny of human-engineered computer viruses. If this eventually happens—if mal-ware entities, through random mutation and natural selection, manage to radically augment their original code and thereby achieve the capacity of sustainable reproduction—then we will witness, for a second time on planet Earth, the mother of all phase changes: the transition from non-life to life.

Soft artificial life research is science, but it's a special kind of science. Whereas hypothesis formation, observation, and experimentation are the mainstays of traditional scientific research, computer simulation is the essential tool used by a-life researchers. Indeed, soft artificial life research was impossible prior to the advent of large-scale, high-speed computers. Repeated iterations of millions of generations of madly replicating, constantly mutating a-life critters is the basic methodology of this research, described by Mark Bedau in one of the standard guides to the field:

> An example of an organismic level artificial life system is Tierra.... This ALife system consists of "organisms" that are actually simple, self-replicating computer programs populating an environment consisting of computer memory and consuming CPU time as a resource. A Tierran genotype consists of a string of machine code, and each Tierran creature is a token of a Tierran genotype. A simulation starts when computer memory is inoculated with a single self-replicating program, the ancestor, which is then left to self-replicate on its own. The ancestor and its descendants repeatedly replicate, until the available memory space is teeming with creatures that all share the same ancestral genotype. To create space in memory for new descendants, older creatures are continually removed from the system. Errors (mutations) sometimes occurs when a creature replicates, so the population of Tierra creatures evolves by natural selection.[17]

It is this iterative process of simulated Darwinian evolution that can coax a startlingly complex and lifelike virtual ecosystem into existence:

> If a mutation allows a creature to replicate faster, that genotype tends to take over the population. Over time, the ecology of Tierran genotypes becomes remarkably diverse. Quickly reproducing parasites that exploit a host's genetic code evolve, and this prompts the evolution of new creatures that resist the parasites. After millions of CPU cycles, Tierra typically contains many kinds of creatures exhibiting a variety of competitive and cooperative ecological relationships.[18]

What is especially surprising is that, just as in natural biological evolution, the evolution of artificial software life inside a computer's memory yields, in Darwin's famous phrase, "endless forms most beautiful and most wonderful" whose emergence are impossible to predict in advance. And, just as with biological evolution, these forms emerge, almost miraculously, from utterly simple beginnings.

Hard Artificial Life

Some a-life theorists believe that life cannot be completely ethereal—that it cannot consist solely of pure information. For these skeptics, the existence of life is inextricably linked to the presence of a material substrate. Real life and real evolution, from their viewpoint, can only occur in a real world composed of interacting aggregations of atoms. Their approach to artificial life consists of setting in motion processes akin to natural selection in evolvable hardware systems. Such systems can consist of computer chips or electric circuits that are capable of reconfiguring themselves, as well as robotic implements that reproduce at differing rates under the pressure of natural selection.

This branch of artificial life research has made relatively little progress, in part because of the technical difficulty of engineering self-reproducing machines capable of evolutionary development through natural selection. That situation may change dramatically with rapid advances in the field of nanotechnology. As imagined by Michael Crichton in his novel *Prey*, self-replicating nanobots (molecule-sized machines invisible to the naked eye) subjected to natural selection could conceivably acquire a frighteningly lifelike ability to evolve at breakneck speed, slipping the leash of human control, and turning on their hapless human creators.

Michael Crichton's *Prey*

The underlying scientific developments that furnish the basis for Michael Crichton's novel *Prey* are nanotechnology (precision engineering at the molecular level) and artificial life. These fields of research have generated dire warnings from the likes of Bill Joy (co-founder of Sun Microsystems) and Astronomer Royal Sir Martin Rees. Joy, for instance, cautions that self-replicating nanodevices, only a few molecules in volume, could conceivably infect and fatally degrade our entire technological infrastructure, transforming everything they touch into "grey goo"—and that no power on Earth would be able to stop the tiny machines once they began reproducing.

In Crichton's hands, this horrifying possibility comes to life. The plot is centered on a high-tech company—Xymos Corporation—which has contracted with the Pentagon to produce a radically new kind of self-propelled spy camera so tiny that it is invisible to the naked eye. What's more, the microscopic devices are capable of communicating and collaborating with one another to produce photographic images of unprecedented clarity.

The problem comes when the devices escape the confines of the Nevada fabrication facility where they are being manufactured and acquire, quite literally, a mind of their own. Predatory swarms of the nanomachines begin reproducing, and they grow more powerful and intelligent by the hour. A computer program called PREDPREY, built into the gadgets to give them a simulated instinct for aggression, mutates into the real thing—and the human creators of the minuscule robots become their prey.

Even more disturbing, the fast-evolving machines figure out a way to infiltrate human hosts and transform them into super-strong and intelligent zombies, easily capable of overpowering and out-thinking uninfected people.

It's a classic Crichton story line: Humans use advanced technology to try to enrich themselves, but end up dead when their creations get loose and violently bite back. Unfortunately, Crichton's somber warning of a technological threat may be all too credible.

Computer guru Ray Kurzweil has predicted that before the 21st century ends, thinking machines will have raced far ahead of humanity in terms of sheer mental ability. When that occurs, the thoughts of the machines may become not merely uncontrollable but literally incomprehensible to us. As astrophysicist Freeman Dyson puts it, the cogitations of these ultra-powerful machine-minds will be as inaccessible to us as our thoughts are to an earthworm or a butterfly. This is the disquieting specter of artificial intelligence research succeeding beyond our wildest dreams (or nightmares).

As Crichton chillingly demonstrates, fast-moving research in nanotechnology and artificial life technologies, some of it funded by the military, raises an even creepier possibility: the creation of mechanical devices no larger than bacteria that are endowed with the ability to reproduce and evolve, and that might eventually come to regard their human creators as nothing more than a tasty lunch or a collection of mobile batteries.

Wet Artificial Life

Wet artificial life is life as we know it—well, sort of. It is life composed of standard-issue biological ingredients such as proteins and information-bearing nucleic acids, very similar to the original version that crawls, slithers, swims, and flies throughout the terrestrial biosphere. But there can be subtle differences.

For instance, wet artificial life might be composed of proteins that string together amino acids with a chirality (or handedness, as in left-handed or right-handed) that is the opposite of that of the amino acids contained in naturally occurring proteins. All living creatures on Earth contain proteins composed of L-amino acids (think of them as left-handed amino acids). Just why this is so is a bit of a mystery, except that it appears to be essential that the proteins in our body be either all left-handed or all right-handed so that they can work together properly. As DNA pioneer Francis Crick pointed out, "Trouble would arise...if you tried to combine the two [protein types], using some components from one system mixed with others from the mirror [proteins]. We can thus see why, in a single organism, the handedness of the many...molecules, large and small, must be concordant."[19] But it appears perfectly reasonable to suppose that the particular choice made by earthly organisms to use left-handed amino acids exclusively as cellular building blocks was purely arbitrary—and that a-life experimenters might conceivably be able to coax life into existence in the laboratory that would be built solely of right-handed amino acids.

Such life would be deeply alien—a kind of mirror life whose biochemical processes, while otherwise identical to our own, operate on a physical substrate that is oriented 180 degrees from the biologically important chemicals in the body of every living creature on Earth.

Similarly, wet a-life researchers might be able to artificially create living creatures that are built of amino acids other than the 20 particular amino acids that are employed to construct proteins by every living creature on Earth. Additional amino acids exist on Earth—indeed, they exist inside the cells of our bodies—but they are never, ever used by earthly creatures as the building blocks for proteins.

Again, this appears to be an arbitrary choice. On other planets, life might have evolved that was built up from an entirely different alphabet of amino acids. One of the goals of wet a-life researchers is to construct living creatures composed of non-standard amino acids. If they succeed, the creatures that emerge from experimenters' test tubes will be as alien as any we might find on a planet orbiting a distant star.

Finally, wet a-life scientists hope to create artificial life that relies on a genetic code that differs from the standard variety employed by all life on Earth. There is no reason to think that the earthly code is the only possible way of creating a set of biochemical instructions for building a living creature. The uniformity of DNA coding probably reflects the unity of origin and relatedness of all terrestrial life and its radiation from a single breeding population of ancestral bacteria. As Francis Crick puts it:

> Even if there existed an entirely separate form of life elsewhere, also based on nucleic acids and protein, I can see no good reason why the genetic code should be exactly the same there as it is here.... If this appearance of arbitrariness in the genetic code is sustained, we can only conclude...that all life on earth arose from one very primitive population which first used it to control the flow of chemical information from the nucleic acid language to the protein language.[20]

There are two main strategies on which wet a-life researchers rely in attempting to build new tailor-made organisms composed of proteins and DNA: a top-down approach and a bottom-up approach.

The Top-Down Approach

The top-down approach to creating artificial life consists of physically extracting parts of the genome of an existing organism and seeing whether the reduced genome can still support living processes. One such experiment is being conducted by human genome pioneer Craig Ventner, who is attempting to prune down the genome of the simplest living cell with the smallest known genome. Ventner and his colleagues are systematically pulling individual genes out of the creature's genome to determine the threshold at which it can no longer reproduce or metabolize. (It sounds a little like pulling the wings off of a fly, but the extraction process is painless for the organism, whose scientific name is *Mycoplasma genitalium* and which lives in the human genital tract.)

A variation of the top-down approach is to insert exotic DNA elements into existing genomes in order to either inculcate desired characteristics in the target organism (the ability to rapidly metabolize oil slicks, for instance) or to simply see what happens when the reengineered gene is expressed in the genetically modified organism.

The top-down approach offers the best prospect for near-term success because it incorporates wholesale the biochemical processes and genetic codes bequeathed to us by evolution, and then tweaks them for experimental purposes. The corresponding disadvantage is that top-down wet a-life researchers are basically dealing with life as we know it, fraught with historical contingency and rife with a hodgepodge of frozen biochemical accidents. The top-down approach doesn't really force researchers to think about the nature and essence of life in a profound way. It's more like re-engineering a Boeing jetliner than uncovering the fundamental principles of physics that make heavier-then-air flight possible.

The Bottom-Up Approach

The bottom-up approach to wet a-life research is entirely different. The goal here is to assemble a living cell from scratch out of non-living organic and inorganic ingredients. This is obviously a more daunting task than the top-down approach, which begins with already living organisms and simply pares them down, gene by gene, to the bare essentials, or inserts genes that were not found in the original creature. The bottom-up approach really is, in a manner of speaking, an attempt to play God—to animate dead matter and endow it with the capacity to metabolize and to reproduce.

No one has been able to pull off this feat to date. As Mark Bedau says, "The main challenge of this bottom-up strategy is that there is no known chemical path for the self-sustenance and reproduction of RNA [ribonucleic acid, a chemical cousin and helpmate of DNA that acts like a kind of chemical translator inside living cells] that is sufficiently complex to encode the minimal molecular functions needed by an artificial cell."[21] Nonetheless, wet a-life researchers remain enthusiastic and optimistic about the bottom-up approach. Enthusiastic—because the creation of life *ex nihilo* would mean that mankind is a giant step closer to answering the ancient question, "What is life?" Optimistic—because once upon a time, long ago, nature pulled off the seeming miracle of transforming dead matter into living flesh. If nature is capable of such a dazzling parlor trick, why isn't humanity?

Strong Artificial Life

An ultimate question confronts researchers working in a-life research: Could real life ever actually be created in a computer memory, in a robotic device, or from a bottom-up assembly of non-living chemicals? This is an issue with ethical and philosophical, as well as strictly scientific, dimensions.

Some researchers are leaning toward a positive answer, pointing to striking parallels between the growth of complexity and organization in the terrestrial biosphere over eons of geological time and a similar but much higher-velocity phenomenon in the so-called technosphere, comprising increasingly powerful Internet-linked computers. Others are more skeptical, arguing that the most a-life researchers can hope to accomplish is to model living processes, not actually create (or re-create) them. To confuse the model with the real thing is to commit what philosophers call a category mistake, analogous to mistaking a computer model of a hurricane with the hurricane itself.

Many a-life scientists believe that this is too facile a response. Soft artificial life experiments such as Conway's Game of Life do not merely model living processes, they contend; rather they actually generate—through natural selection-driven evolution—lifelike phenomena closely resembling metabolism and reproduction. There is no reason apart from "protein chauvinism," these scientists argue, to insist that real artificial life could not emerge in a computer's memory or on the untamed virtual frontier of the Internet.

The Black Cloud

In *The Black Cloud*[22] British cosmologist Fred Hoyle provided a fictional account of humanity's encounter with a highly advanced life form that inhabited an interstellar dust cloud. As the cloud approached our solar system (with the objective of feeding on the energy produced by the sun), it gradually became apparent to astronomers that what initially seemed to be a blob of inanimate matter was, in fact, under intelligent control. Seeking to avert the catastrophe to earthly life that would result from the cloud's blockage of life-giving sunlight, scientists beam messages to the intelligent creature inhabiting the cloud and eventually succeed in establishing two-way communication.

The novel, which was Hoyle's imaginative attempt to argue that life and intelligence were substrate-neutral and could, in fact, thrive in alien environments completely different from the terrestrial biosphere, eerily anticipates the contemporary debate about whether genuine artificial life could emerge inside a computer or network of computers.

The novel also contains a fictional rejoinder to opponents of Hoyle's steady-state theory of cosmogenesis. Queried about how intelligent clouds first emerged and how they first developed defensive screens to shield themselves from radioactive poisoning, the cloud and the human scientists interrogating it exchange the following banter:

> A SCIENTIST: And was your birth, your origin, that is to say, a consequence of spontaneous chemical action, as we believe life here on Earth to have been?
>
> THE CLOUD: No, it was not. As we travel around the Galaxy, we keep a look-out for suitable aggregations of material, suitable clouds in which we can plant life. We do this in rather the way that you might grow saplings from a tree. If I, for instance, were to find a suitable cloud not already endowed with life I would plant a comparatively simple neurological structure within it. This would be a structure that I myself had built, a part of myself. The multitude of hazards with which the spontaneous origin of intelligent life is faced is

> overcome by this practice. Radioactive materials must be rigorously excluded from my nervous system for a reason that I explained in an earlier conversation. To ensure that this is so I possess an elaborate electromagnetic screen that serves to prevent the ingress of any radioactive gas into my neurological regions—into my brain in other words. Should this screen fail to operate, I would experience great pain and would soon die. A screen-failure is one of the possible accidents I mentioned a little while ago. The point of this example is that we can provide our "infants" both with screens and with the intelligence to operate them, whereas it would be most improbable that such screens would develop in the course of a spontaneous origin of life.
>
> A SCIENTIST: But it must have happened when the first member of your species arose?
>
> THE CLOUD: I would not agree that there ever was a "first" member.

Upon hearing this remark from the super-intelligent cloud, the scientists "exchanged a glance as if to say: 'Oh-ho, there we go. That's one in the eyes for the exploding-universe boys.'"[23]

Unfortunately for Hoyle, no intelligent cloud ever materialized in real life to provide proof of his steady-state cosmology. On the contrary, the heavens proclaimed—in the form of the cosmic background radiation—seemingly irrefutable confirmation of the theories of the exploding-universe boys!

So What *Is* Life, After All?

We end this chapter as we began it—by pondering the ancient question: "What is life?" As it turns out, the exercise of examining the state of artificial life research has revealed new intriguing possibilities with potentially revolutionary implications. Remember the hint earlier that life might be synonymous with the evolutionary process? This idea—tendered by one of the leading thinkers in the a-life community, Mark Bedau, and highly controversial, even among this group of thinkers—is that "an automatic and continually creative evolutionary process of adapting to changing environments is the primary form of life."[24]

The Philosophy of Artificial Life

Mark Bedau is a philosopher, not a biologist or computer scientist. What does artificial life have to do with philosophy? According to Bedau, a great deal.

As he explained in a recent paper, a-life and traditional philosophy are natural soulmates, because a-life offers new perspectives on many deep philosophical issues such as the basic nature of life, and whether evolution is totally random or exhibits a kind of robust directionality:

> Philosophy and artificial life are natural intellectual partners, for a variety of reasons. Both seek to understand phenomena at a level of generality that is sufficiently deep to ignore contingencies and reveal essential natures. In addition, by creating wholly new kinds of life-like phenomena, artificial life continually forces us to reexamine what it is to be alive, intelligent, creative, etc. Furthermore, artificial life's computational methodology is a direct and natural extension of philosophy's traditional methodology of *a prior* thought experiment.[25]

The key advantage of artificial life techniques over traditional armchair philosophizing, according to Bedau, is that while "armchair analysis is simply inconclusive," a-life computer modeling imparts a qualitatively different degree of rigor to such traditional philosophical topics as life's essential qualities. As Bedau puts it, "Synthesizing thought experiments on a computer brings a new kind of clarity and constructive evidence to philosophy."[26]

Just to be clear, Bedau is not arguing that evolution *produces* living beings. Rather he is contending that the evolutionary process itself *is* the essence of life:

> Probably the most controversial feature of my theory of life is the claim that supple adaptation does not merely *produce* living entities. The *primary* forms of life are none other than

> the supplely adapting systems themselves. Other living entities are alive by virtue of bearing an appropriate relationship to a supplely adapting system; they are *secondary* forms of life. Different kinds of living entities (organisms, organs, cells, etc.) stand in different kinds of relationships to the supplely adapting system from which their life ultimately derives. In general, these relationships are ways in which the entity is created and sustained by the supplely adapting system.[27]

One of the virtues of this theory, Bedau argues, is that it offers a satisfying explanation of a deep puzzle that has troubled origin-of-life thinkers for centuries: Is there a sharp distinction between living and non-living matter, or is there a shades-of-gray continuum that contains not only life and non-life but also "kinda, sorta" living stuff, such as viruses, prions, frozen sperm, and dormant spores? Bedau comes down squarely in favor of the continuum answer: "If we view life as supple adaptation, then being alive is a matter of degree."[28] *Alive*, one might add, is what the universe gradually becomes as it progresses through intermediate stages of abiotic physical evolution—stages that look an awful lot like evolutionary quantum computation!

Bedau's vision has two dramatic implications regarding the mystery of life's origin: (1) It implies that there is no sharp distinction between prebiotic chemical evolution on the ancient Earth and the subsequent phenomenon of the emergence of evolving life as we know it today; and (2) it implies that the evolving biosphere is the primary manifestation of living matter—Gaia developing over eons of geological time—and that the individual elements of the biosphere (people, insects, bacteria, viruses, and so on) are secondary phenomena, analogous to the cells that constitute our bodies or the mitochondria contained within those cells.

The Emerging Neo-Gaian Synthesis

As religiously motivated Intelligent Design advocates such as William Dembski and Michael Denton attract extensive press coverage for their heated criticisms of Darwinian theory, a far more serious set of challenges to canonical views about the basic mechanism of evolution have begun to emerge beneath the radar of the popular media. A vocal minority of evolutionary theorists have begun to question whether natural selection, acting upon random genetic variation, is adequate to fully explain the observed phenomenon of biological evolution. Darwin himself famously disavowed the assertion that the force of natural selection was, by itself, sufficient to generate the novelty and diversity of the living world. Contemporary complexity theorists such as Stuart Kauffman contend that natural selection is aided by a hidden "handmaiden"—a deep-seated propensity for self-organization embedded in both animate and inanimate nature. Advocates of species cooperation as an evolutionary strategy such as Lynn Margulis likewise stress that natural selection acting on random variation is only part of the saga of the diversification of the biosphere. Contemporary research into the unique role played by viruses in evolution is strengthening the inference that forces beyond selectional pressure play a major role in the unfolding story of life's development and diversification across the planet. My own work suggests that the emergence of increasing levels of bio-complexity may be pre-programmed by the physical constants of nature, which may themselves constitute a novel kind of biosignature.

Collectively, these developments are beginning to coalesce into an alternate vision of evolutionary theory that can supplement the traditional emphasis on Darwinian natural selection, as Darwin himself advocated. This new vision might be provisionally called the emerging neo-Gaian synthesis, in tribute to the pioneering and deeply controversial Gaia theory proposed by James Lovelock. The essence of the emerging neo-Gaian synthesis is that deep symbiosis in the form of species merger and genome fusion serves as a major engine of evolution, accounting not for the *survival* of the fittest—which is indisputably the role of natural selection—but frequently, in De Vries's elegant phrase, for the "*arrival* of the fittest."

In addition, it suggests that the most important feature of future evolution will be its collective character. Bedau's pioneering concept even hints that the biosphere itself may, in the future, exhibit emergent global properties that are not reducible to the behavior or properties of the individual organisms of which it is built—that it will become a true super-organism, alive and hyper-intelligent in strange new ways that its constituent elements (including humans) are incapable of experiencing or even comprehending. That, in essence, is one aspect of the bracing vision of a coming evolutionary singularity.

Chapter 3

Singularities Beyond Singularities

Ray Kurzweil—inventor, scientist, futurist, and best-selling author—can't stop thinking about tomorrow. His prognostications range far and wide, from the future of biotechnology to the prospects for radical life extension, and to the phenomenal technological promise of nanotechnology. But his favorite subject (and the topic of his most recent book) is the looming prospect of a *technological singularity*.

Most scientists associate the term *singularity* with a black hole—a bizarre region predicted by Einstein's theory of general relativity in which an ultra-dense object (such as a collapsed star) bends the fabric of the space-time continuum so severely that nothing—not even a single photon—can escape to the outside universe. But the term has a different, albeit related, meaning in the context of the accelerating progress of technology. In 1958, Los Alamos National Laboratory scientist Stanislaw Ulam reported on a cryptic comment by the great mathematician and computer science theorist John von Neumann that appears to be the first pregnant thought about the implications of a looming technological singularity:

> One conversation centered on the ever accelerating progress of technology and changes in the mode of human life, which gives the appearance of approaching some essential singularity in the history of the race beyond which human affairs, as we know them, could not continue.[1]

Amplifying on von Neumann's casual remark nearly a decade later, the statistician I. J. Good speculated on what the cryptic computer pioneer might have been talking about:

> Let an ultraintelligent machine be defined as a machine that can far surpass all the intellectual activities of any man however clever. Since the design of machines is one of these intellectual activities, an ultraintelligent machine could design even better machines; there would then unquestionably be an "intelligence explosion," and the intelligence of man would be left far behind. Thus the first ultraintelligent machine is the last invention that man need ever make.[2]

The term *singularity* entered the popular science culture with the 1993 presentation at a NASA-sponsored conference of a seminal paper by San Diego State University statistician Vernor Vinge. The abstract of the famous essay is as haunting today as it was more than a decade ago:

> Within thirty years, we will have the technological means to create superhuman intelligence. Shortly after, the human era will be ended. Is such progress avoidable? If not to be avoided, can events be guided so that we may survive? These questions are investigated. Some possible answers (and some further dangers) are presented.[3]

Echoing Good's speculations about the prospects of an intelligence explosion as the essential hallmark of the coming singularity, Vinge went on to draw a scary analogy between this looming technological phenomenon and key patterns discernible in the history of biological evolution:

> What are the consequences of this event? When greater-than-human intelligence drives progress, that progress will be much more rapid. In fact, there seems no reason why progress itself would not involve the creation of still more intelligent entities—on a still-shorter time scale. The best analogy that I see is with the evolutionary past: Animals can adapt to problems and make inventions, but often no faster than natural selection can do its work—the world acts as its own simulator in the case of natural selection. We humans have the ability to internalize the world and conduct "what ifs" in our heads; we can solve many problems thousands of times faster than natural selection. Now by creating the means to execute those simulations at much higher speeds, we are entering a regime as radically different from our own human past as we humans are from the lower animals.[4]

The lessons of our evolutionary past were, in Vinge's view, not exactly comforting:

> From the human point of view this change will be a throwing away of all previous rules, perhaps in the blink of an eye, an exponential runaway beyond any hope of control. Developments that before were thought might only happen in "a million years" (if ever) will likely happen in the next century. [One commentator] paints a picture of the major changes happening in a matter of hours.... [The most disturbing consequence of the technological singularity is that any hyper-intelligent machine] would not be humankind's "tool"—any more than humans are the tools of rabbits or robins or chimpanzees.[5]

Understanding the linkage between our evolutionary past and our probable evolutionary future is of great importance, despite the superficial differences between slow-paced natural biological evolution and hyper-fast technological and cultural natural selection. Indeed, I would hazard a guess that if Charles Darwin were alive today, and fully apprised of the truly revolutionary trends now becoming manifest in what might be called the extended terrestrial biosphere,[6] he would conclude that the sturdy engine of evolution, its vital force undiminished by the passage of centuries, is now poised to hurtle through an invisible barrier and effect a transformational change perhaps equal in import to that ushered in by the Cambrian Explosion half a billion years ago, when multicellular animals, exhibiting a dazzling array of brand-new body plans, began to proliferate in ancient seas.

Darwin would likely conclude as well that *artificial* selection—of which he made artful metaphorical use in *The Origin of Species* to illustrate his hypothesis of speciation through natural selection—has, in our modern era, virtually displaced *natural* selection as evolution's key propellant, at least with respect to the future pathway of human biological development. And the theorist would doubtless contemplate with awe the abiding reality that deep geological time—the enormous stretch of millennia that utterly dwarfs a human lifespan and challenges the very capacity of our biologically evolved human imagination to conceive of its vastness—remains the vital arena in which evolution's epic story continues to unfold.

But the great naturalist would immediately recognize that there is a crucial difference between the process of natural selection as it operated in the distant pass and the novel possibilities currently open to the evolutionary process. A 21st-century version of Charles Darwin would conclude that, though a vision of time's immensity remains the vital key in reaching an understanding of evolution's radical potential, it is a realization of the fathomless magnitude of *future time* and *future history* that is of utmost importance today. A modern Darwin would concur with the conclusion of Princeton physicist John Wheeler: Most of the time available for life and intelligence to achieve their ultimate capabilities lies in the distant cosmic future, not in the cosmic past. As cosmologist Frank Tipler has bluntly stated:

> Almost all of space and time lies in the future. By focusing attention only on the past and present, science has ignored

> almost all of reality. Since the domain of scientific study is the whole of reality, it is about time science decided to study the future evolution of the universe.[7]

Although you won't read about in any *New York Times* or *Wall Street Journal* headline, the disruptive potential of *future evolution* is the emerging leitmotif in advanced biological theorizing today. The current Intelligent Design versus Darwinism dust-up on which the popular press focuses myopically will turn out to be a minor historical footnote to the portentous evolutionary drama that is about to reveal itself in all its unnerving grandeur.

The New Darwin: Ray Kurzweil

Darwin's ghost is unlikely to grace us with a timely appearance, but happily a satisfactory substitute is available: Ray Kurzweil, author of *The Singularity Is Near: When Humans Transcend Biology.* Kurzweil is a computer scientist, prolific inventor, and gifted futurist best known for his two previous bestsellers about artificial intelligence—*The Age of Intelligent Machines* and *The Age of Spiritual Machines.* Bill Gates's enthusiastic blurb for Kurzweil's latest tome proclaims that "Ray Kurzweil is the best person I know at predicting the future of artificial intelligence" and MIT guru Marvin Minsky gushes that this is "a brilliant book with deep insights into the future from one of the leading futurists of our time." This fulsome praise may be on the mark, but it misses the most important point about Kurzweil's current contribution. *The Singularity Is Near* is not primarily a set of predictions about the future of computing or even technology in general. Rather it is a uniquely well-informed, technically literate, and blindingly honest speculation about the very future of evolution itself. *Singularity* is, I submit in all seriousness, the book that Charles Darwin would have written if he were alive today and steeped in the ongoing technological revolution that is engulfing our cultures, our lives, and our minds. As such, it should be required reading for every person with a serious interest in exploring what lies over the horizon in life's ongoing journey from primordial bacterium to transcendent mind.

What's coming next on the evolutionary road, Kurzweil believes, can be inferred from a set of overlapping trends summarized in the 1950s by the legendary information theorist John von Neumann:

> The ever accelerating progress of technology...gives the appearance of approaching some essential singularity in the history of the race beyond which human affairs, as we know them, could not continue.[8]

This scary quotation is, in Kurzweil's view, the key to understanding the future of evolution:

> Von Neumann makes two important observations here: *acceleration* and *singularity*. The first idea is that human progress is exponential...rather than linear.... The second is that exponential growth is seductive, starting out slowly and virtually unnoticeably, but beyond the knee of the curve it turns explosive and profoundly transformative.[9]

What specifically can we anticipate from this process of accelerating change in the near future? What does the coming era of the singularity—defined by Kurzweil as "a future period during which the pace of technological change will be so rapid, its impact so deep, that human life will be irreversibly transformed"[10]—hold in store for the human race?

First and foremost, the fusion of human and machine intelligence. In the post-singularity era, Kurzweil predicts, there will no distinction between human beings and their technologies. As we merge with our machines, we will become something more than merely human. The cyborg-like hybrid entity that is our evolutionary destiny will, in Kurzweil's view, first equal and then transcend the "best of human traits."[11] But, alas, from the viewpoint of garden-variety humans who have not experienced artificial enhancement, this future state of affairs will have dire consequences: it will seem to "rupture the fabric of human history."[12] The only thing that will remain unequivocally human in such a world will be what Kurzweil regards as the defining trait of our humanity: the instinct to exceed our current physical and mental limits.[13]

The limitations that we will transcend will be mental as well as physical. Kurzweil forecasts that we will have personal computer-sized devices capable of emulating human-level intelligence within two decades and effective software models of human thought processes by the mid-2020s. Once machines achieve this level of sophistication and are given the power to improve their own designs, they will inevitably rush past our slower biological brains, achieving mental skills that we can scarcely imagine.

Is this radical vision truly as dystopian as it appears—the end of humanity as we know it? Kurzweil offers a coy answer:

> The intelligence that will emerge [post-Singularity] will continue to represent the human civilization, which is already a human-machine civilization. In other words, future machines will be human, even if they are not biological. This will be the next step in evolution, the next high-level paradigm shift.... Most of the intelligence of our civilization will ultimately be nonbiological. By the end of this century, it will be trillions of trillions of times more powerful than [unenhanced] human intelligence. [14]

Ethics and the Continuum of Intelligence: Crossing the Transhuman Divide

When—not if—a transhuman level of consciousness and intelligence emerges on planet Earth, a pressing moral issue will quickly surface: What ethical and moral obligations will our manifestly superior successors owe to underendowed, hopelessly befuddled *Homo sapiens*?

Kurzweil argues that the probable temporal distance to this disquieting historical discontinuity should be measured in decades, not centuries. Whether the proximate cause turns out to be an exponential quickening of Moore's law (the hypothesis that computer capacity per unit of cost will continue to double roughly every 18 months) or a full-force onslaught of genetic germline engineering (which threatens, according to Princeton evolutionary theorist Lee Silvers, to divide the human race into separate species of unenhanced "Naturals" and genetically augmented "GenRich" individuals) is not really all that significant. What is becoming increasingly clear to thoughtful scientific observers is that, through one path or several, we are about to reach a threshold of radical evolutionary advancement.

What should concern us more urgently with every passing year is the question of what basic canon of ethics and morality, if any, we will succeed in bequeathing to our transhuman successors. In particular, will we be able to endow them with a sense of ethical obligation toward living creatures who are manifestly less intelligent than themselves, perhaps by orders of magnitude? Will we be able to instill in our brainy progeny a principle of fairness and justice that extends across a wide continuum of intelligence and is not cramped and delimited by IQ chauvinism? Will we be able to articulate and embed as a fundamental ethical meme in our successors some principle akin to what I. J. Good called a "Meta-Golden Rule," which Vernor Vinge characterized as a commandment to "treat your inferiors as you would be treated by your superiors."[15]

Based on humanity's own deplorable record with respect to "lower" non-human species, I am not optimistic. As author Douglas Mulhall wrote in 2006:

> To cite a familiar refrain: We are massacring millions of wild animals and destroying their habitat. We keep billions more domestic farm animals under inhumane, painful, plague-breeding conditions in increasingly vast numbers.

The depth and breadth of this suffering is so vast that we often ignore it, perhaps because it is too terrible to contemplate. When it gets too bothersome, we dismiss it as animal rights extremism. Some of us rationalize it by arguing that nature has always extinguished species, so we are only fulfilling that natural role.

But at its core lies a searing truth: our behavior as guardians of less intelligent species, which we know feel pain and suffering, has been and continues to be atrocious.

If this is our attitude toward less intelligent species, why should the attitude of superior intelligence toward us be different? It would be foolish to assume that a more advanced intelligence than our own, whether advanced in all or in only some ways, will behave benevolently toward us once it sees how we treat other species.[16]

Just how far will our nonbiological progeny move beyond those origins? In Kurzweil's exuberant view, very far indeed. In fact, he predicts that the ultimate destiny of brainy thinking machines will be to saturate the entire universe with intelligence:

> In the aftermath of the Singularity, intelligence, derived from its biological origins in human brains and its technological origins in human ingenuity, will begin to saturate the matter and energy in its midst. It will achieve this by reorganizing matter and energy to provide an optimal level of computation...to spread out from its origin on Earth.... [T]he "dumb" matter and mechanisms of the universe will be transformed into exquisitely sublime forms of intelligence.[17]

In short, a capacity to engage in intelligent design of the entire universe is the predicted culmination of the biological and technological evolutionary process that began, so long ago, right here on Earth.

Kurzweil does not, in the know-nothing style of the Intelligent Design community, pose such a capability as a challenge to Darwinian orthodoxy. On the contrary, the ultimate capacity of intelligence is presented as an extension and refinement of Darwin's classic proposition that nature's organized complexity can be explained primarily on the basis of natural selection. According to Kurzweil's worldview, the emergent ability of intelligence to control, configure, and manipulate matter with ever-increasing sophistication is the hypothesized result of an evolutionary process that begins with "dumb" Darwinian natural selection, passes through a technology-creating threshold (which corresponds to our current era), and culminates with the triumph of highly evolved mind over matter and the transcendence of the "dumb" forces of inanimate nature.

Kurzweil's brave vision is in the spirit of Charles Darwin's historic theory—an entirely plausible and utterly iconoclastic analysis of life's seemingly miraculous capacity to yield, in Darwin's unforgettable phrase, "endless forms most beautiful and most wonderful" through fully naturalistic means. The fact that, as did Darwin, Kurzweil tends to demote humanity from the centerpiece of creation to a supporting role in the vast emerging spectacle of the cosmos, is perhaps his greatest intellectual virtue.

Signs of a Secular Rapture

Now that we have a basic understanding of what is meant by a technological singularity (at least as the term is understood by Ray Kurzweil), what are likely to be some of its concrete manifestations? And what will be the telltale signs that we have entered an era that one wag described as "a rapture for the rest of us"?[18]

Life Eternal (or at Least Until You Get Bored Out of Your Mind)

What if you could live forever—or at least as long as you choose? The dream of eternal life has been the nearly exclusive province of religious faith throughout recorded history, but there is now solid evidence that science is about to transgress this sacred enclave. The emergence of realistic technological methods for achieving radical life extension, it is confidently predicted, will be one of the hallmarks of the coming era of the singularity.

According to Kurzweil and Terry Grossman, M.D., coauthors of *Fantastic Voyage: Live Long Enough to Live Forever*,[19] three "bridges" will allow humanity to cheat death and cross over into a gleaming realm of *de facto* immortality. As Kurzweil put it in a recent interview:

> Terry Grossman and I have described what we call the "three bridges" to radical life extension. Bridge one has to do with taking full advantage of today's knowledge of biology in order to dramatically slow down aging and disease processes. This

> will enable us to stay in as good a shape as possible for when bridge-two technologies become available. Bridge two is the biotechnology revolution, which will give us the tools to reprogram our biology and the biochemical information processes underlying our biology. We're in the early stages of that revolution already, but in fifteen years we will have, to a large extent, mastery over our biology. That will take us to the third bridge, the nanotechnology revolution, where we can rebuild our bodies and brains at the molecular level. This will enable us to fix the remaining problems that are difficult to address within the confines of biology and ultimately allow us to go beyond the limitations of biology altogether. So the idea is to get on bridge one now, so we can be alive and healthy when the biotechnology and nanotechnology revolutions come to fruition. Our aim is to live long enough to live forever.[20]

The third bridge sounds reminiscent of a transformation of humanity into a human/machine hybrid—and Kurzweil makes it clear that is exactly what he envisions:

> As we merge with our technology, we will have billions or trillions of nanobots in our bloodstreams keeping us healthy, interacting with our biological neurons, and providing, for example, full-immersion virtual reality incorporating all of the senses. If you want to be in real reality, the nanobots will just sit there and do nothing. If you want to be in virtual reality, they'll shut down the signals coming from your real senses, replace them with the signals that you would be experiencing if you were in the virtual environment, and your brain will feel like it's in that virtual environment. You can move your virtual body there and have any kind of encounter you want, incorporating all of the senses.[21]

The real payoff of this human/machine merger is not the availability, through virtual reality, of "any kind of encounter you want," but rather a qualitative leap upward in cognitive capacity:

> But most importantly, this intimate merger of our biological intelligence with nonbiological intelligence will vastly expand human intelligence as a whole. I mean, once it gets a foothold in our brains, our thinking will really be a hybrid of the two,

> and ultimately, the nonbiological portion will be much more powerful, and may give us access to new forms of intelligence that are very different than anything we've experienced.[22]

Access to this new kind of intelligence will be essential, Kurzweil believes, because without it, we would grow bored out of our minds during the course of a 500- or 1,000-year lifespan:

> Psychologically, we're not equipped to live five hundred years. So if we are talking only about conquering disease and aging, and then just living on as human beings in our current form for hundreds or thousands of years, that would lead to a serious problem. I think we would develop a deep ennui, a sort of profound despair. We would get bored with the level of intelligence we have and the level of experience we have available to us. I think in order to make this viable, we need not only radical life extension but radical life expansion. We need to expand our intelligence and our capacity for experience as well, which is exactly what these new technologies will enable us to do. Then an extended life span would become not only tolerable but a remarkable frontier where we could pursue the real purpose of life, which is the creation and appreciation of knowledge.[23]

Not everyone agrees that radical life extension would be an unalloyed benefit to mankind. One of the most articulate critics (of what she regards as a thoroughly dystopian state of affairs) is evolutionary biologist and ecologist Connie Barlow. In a recent interview she expressed distaste for the very idea of radical life extension for three distinct reasons:

> I don't like the prospect. For one thing, it will exacerbate the schism between the haves and the have-nots because, obviously, the whole world isn't going to have access to this. For another, I view it as undesirable because we're having enough trouble right now limiting our reproduction, and if we have a significant number of people who are engaged in that sort of life extension, it will create even more of a population problem on earth. But more fundamentally, I think that our tendency to avoid the thought of death or think that there's something wrong with death actually limits our understanding of life and our zest for life.[24]

Indeed, Barlow makes a convincing case that the complexification of the biosphere through biological evolution and, beyond that, the very creation of heavier elements such as carbon and oxygen (discussed in the introduction) through the stellar alchemy that takes place in exploding supernovae provides indisputable evidence that death—of individual organisms, of entire species, and of the mightiest of stars—is essential to the appearance of life and to Darwinian natural selection. Indeed, death is the blade wielded by nature to shape and sculpt the living firmament through Darwinian processes; absent death, there could be no such phenomenon as natural selection.

To proclaim an intent to evade death is, in Barlow's view, to commit an unspeakable heresy that offends a primordial principle of nature:

> From the smallest levels within our bodies to the largest levels out there in the universe, we have a whole nested reality in which death is not just natural, it's creative. It's what allows everything to be. Were it not for death, there would be no such thing as food. Everything we eat was once alive. When you're eating salad, or anything that's uncooked, those cells are still alive right at the moment you're eating. You're killing them as they go into you. Even if immortality comes about in some way, we still can't eliminate death from the whole cycle of life.[25]

Machines Begetting Machines

The onset of the singularity will, in Kurzweil's view, be driven by three overlapping technological revolutions in genetics, nanotechnology, and robotics, of which the latter will be the most dramatic and consequential. As he puts it:

> The most powerful impending revolution is the robotic revolution. By robotic, I am not referring exclusively—or even primarily—to humanoid-looking droids that take up physical space, but rather to artificial intelligence in all its variations.[256]

This development will result in the appearance of a form of *strong artificial intelligence* that will, for the first time, supplant the human mind as the dominant form of intelligence on the planet:

> By the end of this century, computational or mechanical intelligence will be trillions of trillions of times more powerful than unaided human brain power.... Artificial intelligence will necessarily exceed human intelligence for several reasons. First,

> machines can share knowledge and communicate with one another far more efficiently than can humans.... Second, humanity's intellectual skills have developed in ways that have been evolutionarily encouraged in natural environments. These skills, which are primarily based on our abilities to recognize and extract meaning from patterns, enable us to be highly proficient in certain tasks, such as distinguishing faces, identifying objects, and recognizing language sounds. Unfortunately, our brains are less well-suited for dealing with more-complex patterns, such as those that exist in financial, scientific, or product data.... Finally, as human knowledge migrates to the Web, machines will demonstrate increased proficiency in reading, understanding, and synthesizing all human-machine information.[27]

Despite their daunting intellectual superiority, these hyper-intelligent machines should still be viewed, at least in a certain sense, as human:

> I argue that computer, or as I call it *nonbiological intelligence*, should still be considered human since it is fully derived from human-machine civilization and will be based, at least in part, on a human-made version of a fully functional human brain.[28]

But it seems intuitively obvious that the way to truly supercharge the evolution of machine intelligence is to gradually ease human beings out of the design loop so that machines are placed in charge of the design of new generations of machines. It seems equally obvious that, should this occur, brainier and brainier cohorts of artificial intelligences will gradually drift away from the human mold, becoming more and more alien with each passing year.

This prospect seems to have occurred to Kurzweil, who believes that a "narrow relinquishment of the development of certain capabilities needs to be part of our ethical response to the dangers of twenty-first century technological challenges."[29] In particular, he applauds the development of voluntary safeguards in the field of nanotechnology:

> Another constructive example of this are the ethical guidelines proposed by the Foresight Institute: namely, that nanotechnologists agree to relinquish the development of physical entities that can self-replicate in a natural environment free of any human control or override mechanism.[30]

However—and somewhat paradoxically—Kurzweil seems sanguine at the prospect of "runaway AI": the exponential acceleration in AI capabilities that is likely to ensue when machine intelligences gain access to their own design specifications and are able to directly intervene in their evolutionary future by engineering improvements in their progeny:

> The logic of runaway AI is valid, but we still need to consider the timing. Achieving human levels in a machine will not *immediately* cause a runaway phenomenon.... It will take time for computers to master all of the requisite skills.... [T]he extraordinary expansion contemplated for the Singularity, in which human intelligence is multiplied by billions, won't take place until the mid-2040s.[31]

Machine intelligence won't exceed the human variety *by a billion-fold* until the mid-2040s. So what have we happy-go-lucky humans got to worry about? Humanity can look forward to a stretch of around 30 years from the moment you read this sentence until the day arrives—think of it as Singularity Judgment Day—when artificial forms of intelligence gain control of their own destiny and race past the pitifully outclassed computers housed in the skulls of *Homo sapien*!

The Turing Test

The Turing test, named after its inventor Alan Turing, provides a simple criterion for deciding whether a machine has acquired human-level intelligence: Can the machine converse so naturally in an open-ended conversation with an average human being that it could fool the human into believing that it was, in fact, another person? As speech-recognition technology has progressed, some experts have begun to believe that the Turing test is insufficiently demanding and propose a more exacting standard—such as the ability of an artificial intelligence to converse knowledgeably with a peer review group of scientists and convince the scientists that it was another human expert in their field. The irony of this variation on the Turing test is that not many flesh-and-blood humans would be able to pass it!

The Emergence of Collective Consciousness

Perhaps the emergence of superintelligence will come about, not through the design and fabrication of ever more powerful computers, but as a result of collective efforts—maybe even *unconscious* collaborations—of human and machine nervous systems linked together in a highly evolved version of today's Internet. Perhaps such a collaborative project could transmute what *New Yorker* columnist James Surowiecki calls "the wisdom of crowds"[32] into a transhuman capacity for collective analysis and decision-making that would dwarf the formidable predictive capacities of human knowledge markets such as stock exchanges and the Iowa Electronics Markets (IEM), which traffic in political futures.

The Uncanny Accuracy of the IEM and the Possibility of Market-Based Terrorism Prediction

The Iowa Electronics Markets (IEM) project was created in 1988 by the College of Business at the University of Iowa on the basis of a daring conjecture: Futures markets might offer a better way to predict the outcome of contested political elections than conventional polls. The idea was that futures markets are superb integrators and aggregators of the hunches, intuitions, and educated guesses of a myriad of market participants with a financial incentive to place their bets smartly. As James Surowiecki wrote in *The Wisdom of Crowds*:

> EM features a host of markets designed to predict the outcomes.... Open to anyone who wants to participate, the IEM allows people to buy and sell futures 'contracts' based on how they think a given candidate will do in an upcoming election.[33]

The track record of the IEM is nothing short of stunning. As Surowiecki notes:

> The IEM has generally outperformed the major national polls, and has been more accurate than those polls even months in advance of the actual election.[34]

After the terror attacks of September 11, 2001, in New York City and Washington, DC, the IEM's extraordinary success at predicting outcomes in uncertain situations caught the attention of the Pentagon. Defense intelligence officials were concerned that no efficient means existed of aggregating intelligence estimates made by an alphabet-soup of intelligence agencies and, more important, of coming up with consistently reliable judgments about what the raw data meant in terms of the probability of future terrorism incidents. Perhaps, the DOD speculated, the IEM approach could be modified to improve the federal government's terrorism threat assessment procedures.

Thus was born a unique DARPA project. (DARPA is the acronym for the Defense Advanced Research Projects Agency, a uniquely nimble federal agency that has been directly responsible for such technological innovations as the Internet.) The DARPA project was called FutureMAP; it incorporated an element copied directly from the IEM called the Policy Analysis

Market (PAM). The PAM was intended to be a market in terrorism futures. Members of the public were to be invited to bet on the likelihood of possible terrorism events, with the objective of dramatically improving the predictability of such threats in advance of another catastrophe.

As soon as it became public, the DARPA project generated a firestorm of political criticism on Capitol Hill. Senator Ron Wyden of Oregon led the charge to kill the innovative program, labeling it harebrained, offensive, and useless. As Surowiecki concludes, Wyden's politically motivated broadside against PAM may have seriously weakened national security.[35]

This human *intelligence amplification* (or IA) approach, rather than an exponential growth in raw computer processing capacity, is the technological superhighway that will lead most quickly to greater-than-human intelligence, at least in the view of Vernor Vinge:

> When people speak of creating superhumanly intelligent beings, they are usually imagining an AI project. But...there are other paths to superhumanity. Computer networks and human-computer interfaces seem more mundane than AI, and yet they could lead to the Singularity. I call this contrasting approach Intelligence Amplification (IA). IA is something that is proceeding very naturally, in most cases not even recognized by its developers for what it is. But every time our ability to access information and to communicate it to others is improved, in some sense we have achieved an increase over natural intelligence. Even now, the team of Ph.D. humans and good computer workstations (even an off-net workstation!) could probably max any written intelligence test in existence.[36]

The reason this approach may offer a shortcut to superhuman intelligence is that the constituent parts of the collective computing enterprise (including fully functioning assemblies of biological neurons known as human brains) are available off the shelf and in virtually limitless supply. As Vinge puts it:

> [I]t's very likely that IA is a much easier road to the achievement of superhumanity than pure AI. In humans, the hardest development problems have already been solved. Building up from within ourselves ought to be easier than figuring out first what we really are and then building machines that are all of that. And there is at least conjectural precedent for this

> approach. Cairns-Smith has speculated that biological life may have begun as an adjunct to still more primitive life based on crystalline growth. Lynn Margulis...has made strong arguments that mutualism is a great driving force in evolution.[37]

What will be the catalysts that propel the emergence of collective human/machine superintelligence? At the 2005 Accelerating Change Conference on AI and IA, two key enabling technologies and developments were identified. The first was a fully functional *conversational user interface* by means of which humans could converse, more or less in natural language, with their computers:

> Achieving a functional conversational user interface (CUI) would be perhaps the single most important and empowering artificial intelligence/intelligence amplification breakthrough we may witness in our lifetimes. It would give us the ability to talk to, be productive with, and be continually educated by our computers, cellphones, the Internet, and other complex technologies using simple but natural human conversation. Moving beyond today's early voice response and language processing systems, the first reasonably sophisticated CUIs will allow us to converse semi-naturally on an ever-growing range of topics with our machines, and to develop a level of personalization and sophistication in our public and private preferences, user histories, networking and knowledge and relationship management systems that is presently unattainable.[38]

A second catalytic trend, identified at the conference as an IA enabler, was the proliferation of what panelists called a virtual metaverse—an ever-expanding ensemble of online 3D worlds where a growing number of real human beings spend most of their waking hours, creating copyrightable original materials and patentable inventions, thereby earning their real-life livings while immersed in a dazzling array of diverse computer-generated virtual realities.[39] That particular pathway to the singularity at least promises some fun along the way!

Artificial Mirror Neurons: Key to Genuine AI?

In *The Singularity Is Near*, computer science guru Ray Kurzweil argues that a promising strategy for achieving the goal of genuine artificial intelligence is to reverse-engineer the human brain. This task is already underway at leading research institutions around the world. As reported in 2006 in *Scientific American*:

> Recent technological advances are narrowing the gap between human brains and circuitry. At Stanford University, bioengineers are replicating the complicated parallel processing of neural networks on microchips.[40]

Top brain scientists now believe that a key driver of the evolution of human-level intelligence was the proliferation of mirror neurons in early humans.[41] In the view of one leading researcher, V. S. Ramachandran of the University of California, San Diego, mirror neurons may hold the key to understanding how and why humanity took a "great leap forward" in cognitive power approximately 50,000 years ago, acquiring cultural skills unique to humans that made possible the appearance of language, tool use, art, and eventually science and technology:

> The discovery of mirror neurons...and their potential relevance to human brain evolution...is the single most important "unreported" (or at least, unpublicized) story of the decade. I predict that mirror neurons will do for psychology what DNA did for biology: they will provide a unifying framework and help explain a host of mental abilities that have hitherto remained mysterious and inaccessible to experiments.[42]

Mirror neurons, as the term suggests, are specialized brain cells that facilitate imitation and empathy. They fire not only when we perform a certain activity but also when we observe that activity being performed by others. As a recent *Scientific American* article put it:

> The discovery of this mechanism...suggests that everything we watch someone else do, we do as well—in our minds. At its most basic, this finding means we mentally rehearse or imitate every action we witness, whether it is a somersault or a subtle smile. It explains much about how we learn to smile, talk, walk, dance or play tennis. At a deeper level, it suggests a biological dynamic for our understanding of others, the complex exchange of

ideas we call culture, and psychological dysfunctions ranging from lack of empathy to autism.[43]

If Ramachandran is correct in hypothesizing that mirror neurons are responsible for the so-called "cultural Big Bang" that occurred about 50,000 years ago and that started humanity on the pathway toward language, tool use, and technological civilization, then it seems apparent that a promising strategy for AI researchers would be to reverse-engineer human mirror neurons and insert them into silicon-based neural networks. In addition, it appears equally obvious that, in order to harness their unique capabilities, researchers should not only insert artificial mirror neurons into an artificial neural network but should also provide them with access to external sensory stimuli. This notion reiterates one of the most daring and interesting suggestions of computer pioneer Alan Turing in 1948: A robot that scientists might hope to imbue with genuine artificial intelligence "should be allowed to 'roam the countryside' so that it would be able to 'have a chance of finding things out for itself.'"[44]

The Future of History: Just One Singularity After Another

In 1992, Francis Fukuyama famously erred by proclaiming the end of history and the arrival on the global scene of the "last man." Written soon after the fall of the Berlin Wall, Fukuyama's blissfully optimistic book opined that all of the nations of the world were moving inexorably toward liberal democratic self-governance and free-market capitalism. The contrary views of historian Samuel Huntington, who predicted an ongoing clash of civilizations, turned out to be closer to the mark, but neither scholar foresaw (at least at the time) the massive and disruptive impact of ever-accelerating information technology on the very fabric of human history. (Perhaps this is because both men come from the liberal arts side of the academy.)

It should be clear by now that the future will differ radically from the past; it will be at least as different as the radically new world of biological complexity and diversity ushered in by the Cambrian Explosion was from the preceding era.

The Virtual Cambrian

Approximately half a billion years ago, life crossed a portentous threshold. Over a period of few million years—a fleeting instant in geological terms—a wild profusion of animal body plans appeared. This sudden arrival on the evolutionary stage of prototypes of all currently living multi-cellular animals is known as the *Cambrian Explosion* or the *Cambrian Radiation*.

The Cambrian Explosion was an evolutionary watershed because it opened the floodgates to future evolution. As Richard Dawkins has noted, the creatures of the Cambrian were "champion evolvers," capable of exploiting and creating ecological niches that had not previously existed on planet Earth. One important consequence of that evolutionary spurt was the steady growth, over ensuing millennia, in the variety and complexity of multi-cellular animals, culminating in the appearance of primates and eventually modern humans.

When humans arrived on the scene, they brought with them the ability to create a novel kind of natural selection that had not previously existed, at least in any significant way: cultural evolution. This new form of evolution led to the origin and diversification of languages, the birth of a multitude of religious belief systems, and the eventual dawn of technology and science.

Whereas the units of genetic transmission are called *genes*, the units of transmission responsible for culture evolution are called *memes*.

Cultural evolution bears some similarities to biological evolution, but a key difference is that cultural evolution is immensely faster.

In the late 20th century, the forces of cultural evolution crossed a crucial threshold: They acquired the technological capability to intervene directly in the redesign of the human genome (as well as the redesign of the genomes of other living creatures). This new technology is called germline genetic engineering. As a result, life on planet Earth has once again passed an evolutionary watershed, equal in eventual magnitude to the Cambrian Explosion half a billion years ago.

I call this new threshold—this new evolutionary watershed—the Virtual Cambrian. The dawn of the Virtual Cambrian means that many momentous changes await us on the other side of this historical event horizon, including: (1) a vast acceleration in the pace of future biological evolution, because the pacemaker for this new kind of deliberately engineered evolutionary change will not be the slow tick-tick-tick of random genetic mutation (the engine of classic Darwinian evolution), but lightning-fast cultural drivers such as changing trends in body-style fashions (will we want our offspring to be tall or short, stout or thin, super-strong or super-graceful?) and changing preferences on the part of successive generations of parents regarding the optimal mental, social, and athletic proclivities of their offspring; (2) a rapidly spreading wave of empowerment on the part of prospective parents to actually shop in what one scientist called the "genetic supermarket" by tailoring an embryo's genome to increasingly precise design specifications (projected age span of 200 years, blue eyes, red hair, fair skin, male, homosexual, 6 feet tall, IQ of at least 180, and freedom from all identifiable predispositions to genetic diseases); and (3) the predictable result will be a runaway explosion in gene/culture coevolution—a kind of arms race on the part of parents to bio-engineer better and better babies—that will make the frantic competition on the part of ambitious parents to get their children into the best colleges look like, well, child's play.

With the dawn of the age of the Virtual Cambrian we seem destined, as was Prometheus, to seize the inner fire of evolution and transform it into a utilitarian tool. The great unknown is the precise path that gene/culture coevolution will take in the near and distant future. But what can be stated with certainty is that, Faustian trepidations notwithstanding, the looming gene/culture coevolutionary process enabled by germline genetic engineering will be irresistibly powerful and unnervingly fast.

We can no more hope to resist the force of the Virtual Cambrian explosion that awaits us just over history's frontier than the simple one-celled organism that are our own distant ancestors could have hoped to stem the flood of biological innovation unloosed by the first Cambrian Explosion 500 million years ago.

The only real question is whether the creatures (biological and artificial) that will emerge from the Virtual Cambrian will bear any recognizable kinship—any hint of spiritual or biological consanguinity—to human beings.

What may not be so obvious is that the singularity will not be a singular event. There will likely be multiple singularities, succeeding one another with accelerating rapidity. To use an analogy, it will be as if there are an infinite succession of black holes nested like babushka dolls inside of other black holes, each more wrenching and disruptive than the last.

Superhuman artificial intelligence will be no more immune than human intelligence to the ensuing historical discontinuities. If humans will be first to be tossed off the throne as the intellectual monarchs of the planet, our computer-based AI successors won't be far behind us, succeeded by...who knows what? Forms of intelligence our puny human brains cannot even begin to imagine? Or maybe descendants of intelligent creatures that acquired sentience billions of years before humankind's ancestors descended from the trees? I am speaking, of course, of extraterrestrial intelligence.

The Intelligent Universe

Part II

Extraterrestrials

In Part I of this book, we toured the far-flung precincts of the computational universe, encountering such marvels as software that writes itself through a life-like process of simulated Darwinian evolution, bits and bytes of data striving to come to artificial life in the innards of a computer, and the dizzying prospect of the singularity—a rapidly approaching historical inflection point at which technological progress will become so rapid that our human minds will be left hopelessly behind by ever-improving robotic intellects.

Reaching the singularity will mean that evolution will achieve a kind of escape velocity, allowing it to transcend the legacy of past eons of glacially paced trial-and-error natural selection. The vision of the singularity provides an appropriate transition to Part II of this book, in which we will shake off the surly bonds of our home planet and search the stars for signs and portents of extraterrestrial intelligence. In the process, we shall discover a unique—but realistic—method by which our distant virtual progeny might zip across the galaxy at the speed of light. Finally, we will consider the outlandish possibility that we might, in the distant future, be capable of creating new baby universes by deliberating setting off new Big Bangs, and then achieving communications of sorts with the Big Babies by carefully engineering the rules of physics that will govern their future evolution.

The Intelligent Universe

Chapter 4

The Fermi Paradox Revisited

In my home state of Oregon, famous for its forests of Douglas fir and for an unspoiled coastline, lies the mighty Cascade mountain range. The best known peak in the range is the active volcano Mount St. Helens, located just north of Portland in the state of Washington. But many Oregonians would agree that the most beautiful peaks in the Cascade range are a trio of mountains in central Oregon known as the Three Sisters. The names given to these three peaks were originally chosen by members of the Methodist Mission in Salem, Oregon, in the early 1840s. The individual mountains were designated Mount Faith, Mount Hope, and Mount Charity.

As we prepare to embark on a new age of space exploration at least as daunting as the era of epic migration experienced by Oregon's pioneers when they crossed the Cascade mountains, it is tempting to think of the three planetary bodies in our solar system most congenial to our kind of carbon-based life as three planetary sisters. Born of the same massive nebular cloud that also birthed the giant outer gas planets (Jupiter, Saturn, Neptune, and Uranus) as well as the life-sustaining sun, the planetary sisters are guardians of the deepest and most ancient of cosmic mysteries—why there exists a pathway, still largely shrouded from human comprehension, by which inanimate matter transforms itself over eons of geological time into vibrant, self-renewing, ever-evolving life.

First, of course, is the innermost sister: Earth, lush with a spectacular biosphere so robust that it has reshaped the very air, land, and sea of the entire planetary surface. Think of our home as Planet Faith. By virtue of her sheer biological exuberance, the example of Earth inspires us to have faith that living worlds may be abundant throughout the cosmos.

Think of the second planetary sister—rusty, dusty Mars—as Planet Hope. We are now receiving a steady stream of tantalizing scientific evidence that Mars may once have been covered with liquid water and a thick atmosphere. New findings reveal that vast frozen oceans are probably still buried beneath the arid Martian surface. More controversially, a chunk of rock found in Antarctica that was blasted from the surface of Mars by an asteroid impact, as well as the recent discovery by the European Space Agency's Mars Express of traces of methane in the Martian atmosphere, hint that life may continue to exist somewhere below the plains and dry river beds of Mars. (Chemically, methane is out of place in the carbon dioxide atmosphere of Mars; its presence may be a bio-marker, just as the out-of-equilibrium presence of free oxygen in Earth's atmosphere is a bio-marker, evidencing terrestrial plant life.) Collectively, this evidence leads many scientists to *hope* that life once existed on Mars, and might have endured to this day.

The outermost planetary sister—call her Planet Charity—is not really a planet at all, but the ocean-moon Europa, which orbits Jupiter, gravitational master of the outer solar system. The scientific evidence is now overwhelming that a planetary sea, larger than all of Earth's oceans combined, lies beneath Europa's icy crust (see figure on page 93). This sea seems to be heated, not by the warmth of the sun, but by a kind of incessant kneading performed by the massive gravitational force exerted by Jupiter. (Try to envision a tidal force exponentially more powerful than that exerted by our moon on Earth's oceans.) Here, gratuitously and unexpectedly, is a third potential living world. Located far beyond what scientists normally consider to be the habitable zone of the solar environment (defined to be that area of outer space surrounding the sun where planetary equilibrium temperatures are consistent with surface liquid water), Europa is a fantastically improbable venue that just might host living creatures. Europa's bio-friendliness is the unforeseen bequest of a *charitable* natural order—nature's unanticipated provision of a potential situs for life in a most unlikely location.

But what if life proves to be limited to Earth in our own solar system? Must we then discard faith and hope in the possibility of a cosmos filled with living creatures? Must we conclude that a decidedly uncharitable nature has animated dead matter with the spark of life only once since the Big Bang?

Absolutely not! For far beyond our own humble solar system stretches the endless domain of interstellar space where numberless legions of exo-planets orbit distant starts. (NASA's upcoming Kepler space mission is expected to detect hundreds of exo-planets beyond the hundred-plus giant extrasolar planets currently known to orbit nearby stars.) Surely on one of these planets circling one of the billions of stars in the Milky Way galaxy, the miracle of life's origin and evolution must have been repeated. And if not within our own galaxy, then certainly living creatures must have emerged on a planet circling a star in one of the billions of other galaxies sprinkled throughout the visible universe.

The first Western scholar to articulate scientific speculations about the possibility of life on planets orbiting distant stars was an extraordinary 16th-century visionary named Giordano Bruno. As the Spanish physicist Beatriz Gato-Rivera wrote, "[Bruno] claimed that the sun was only one star among the many thousands, and therefore, like the sun, many other stars would also have planets around them and living beings inhabiting them."[1] Bruno's was a speculative insight of extreme prophetic genius, at the very outer limits of the capacity of the scientific imagination.

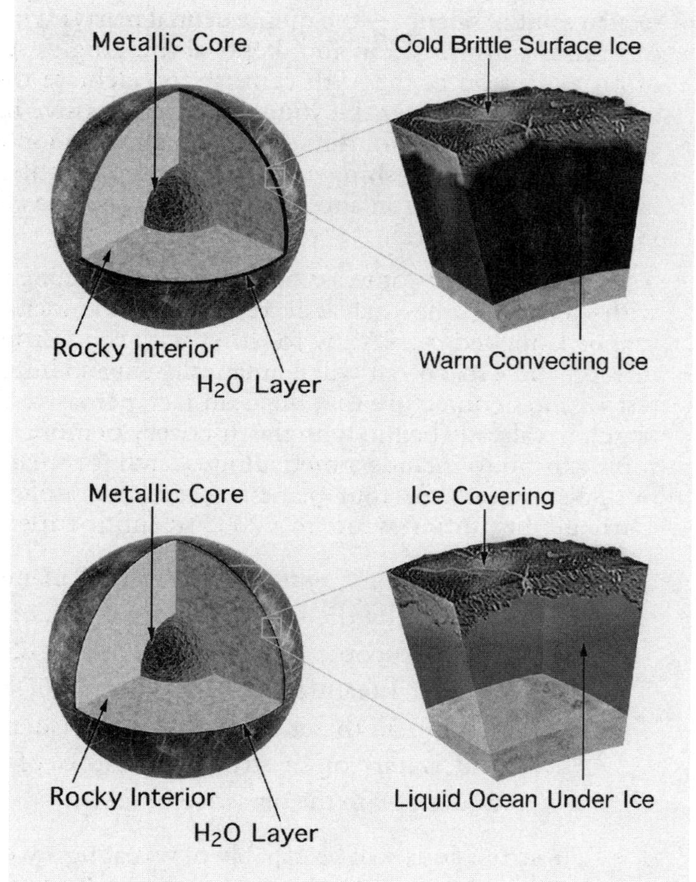

NASA diagram of the composition of Europa
Image provided by NASA

As Gato-Rivera notes:

> To appreciate the genius of Giordano Bruno one has to take into account that he lived at the time when more than 99% of the intellectuals believed that the Earth was the center of the Universe, and a few others, like Copernicus and Galileo, believed that it was the Sun, instead, at the center of the Universe, the stars being some bright heavenly bodies of an unknown nature. Nowadays we know that the Universe has no center and that our planet is only a tiny particle of dust in its immensity.[2]

For these heretical insights, Giordano Bruno was rewarded by the Catholic Church with imprisonment and execution: He was burned at the stake in Campo dei Fiori in Rome on February 17, 1600. Unfortunately, the church (which tried to make amends for its hostility toward Galileo) has never apologized for executing Bruno. Perhaps some heresies are still too dangerous to forgive, despite their scientific merit.

Though Bruno has yet to find favor with the pope, he has become a kind of secular saint of science—the quintessential martyr who died to defend the integrity of scientific investigation and skeptical thinking. A statute was erected on the site of his execution in the 19th century to celebrate the principle of free scientific inquiry. More recently, the youngest of the massive impact craters that pockmark the moon was named for him. And the highest honor bestowed by the SETI League (a leading group pushing to expand the scientific search for extraterrestrial intelligence) is the Giordano Bruno Award, given to celebrate outstanding advances in this field of research.

In the years to come, a whole fleet of astonishing new space-borne laboratories with evocative names such as Terrestrial Planet Finder, Life Finder, and Planet Imager will be launched by NASA. Together with the European Space Agency's Darwin mission, these spacecraft will dramatically expand humanity's ability to scientifically test Bruno's conjecture that life is, in fact, pervasive throughout the cosmos. The search has already begun with the discovery of more than 100 exo-planets, and will soon expand to include a meticulous search for smaller Earth-like planets outside our solar system. As four planetary scientists affiliated with Princeton and the Carnegie Institution wrote in a 2005 scientific paper:

> The search for extrasolar terrestrial planets is in large part motivated by the hope of finding signs of life or habitability via spectroscopic biosignatures. Spectroscopic biosignatures are spectral features that are either indicative of a planetary environment that is hospitable to life (such as the presence of liquid water) or of strong indicators of life itself (such as abundant O_2 in the presence of CH_4).[3]

These missions will be capable of revealing a wealth of information about an alien world orbiting a distant sun "such as the existence of weather, the planet's rotation rate, the presence of large oceans or surface ice, and the existence of seasons."[4] Even more exciting is the probability that these missions will be capable of directly detecting evidence of the presence of life—biosignatures—both in an exo-planet's atmosphere (through the presence of out-of-equilibrium conditions such as the presence of massive quantities of free oxygen) and on its surface (by registering a spectroscopic signature indicating the presence of surface vegetation biochemically similar to terrestrial vegetation).

These missions will even be capable of distinguishing planet-sized *artificial* structures orbiting a distant star from natural celestial bodies. This emerging detection capability opens up a new window on the universe that might conceivably reveal the presence of alien intelligence. As Luc Arnold, a scientist affiliated with the Observatoire de Haute-Provence in France, wrote in a 2005 paper:

> Current Search for Extraterrestrial Intelligence (SETI) programs concentrate on the search for radio or optical laser pulse emissions. We propose here an alternative approach for a new SETI: considering that artificial planet-size bodies may exist around other stars, and that such objects always transit in

> front of their parent star for a given remote observer, we may thus have an opportunity to detect and even characterize them by the transit method, assuming these transits are distinguishable from a simple planetary transit. These objects could be planet-size structures built by advanced civilizations, like very lightweight solar sails or giant very low density structures maybe specially built for the purpose of interstellar communication by transit.[5]

Despite such exciting new theories (and despite the emergence of ever more capable detection methods and space-based observatories), one nagging question must be acknowledged: If life in general—and intelligent life in particular—is pervasive throughout the countless galaxies in our universe, then where is everybody? This is the famous Fermi Paradox, named after physicist Enrico Fermi, who posed the question during a luncheon conversation at the Los Alamos National Laboratory in 1950. This issue has been sharpened in recent years by scientists who point out that because we inhabit a very old cosmos, multitudes of sun-like stars formed billions of years before our sun. If the emergence of life and intelligence is truly preordained by the laws of physics and chemistry, then at least some of those stars should be surrounded by life-friendly planets hosting vibrant biospheres on which intelligent creatures evolved billions of years ahead of mankind. By now, civilizations composed of such creatures should have acquired the technology to conquer and colonize entire galaxies, including our own Milky Way galaxy. However, we have uncovered no credible evidence of their presence. (My sincere apologies to the legions of UFO believers, telepathic contactees, and abduction claimants. Your clamorous assertions that Earth may have been visited by intelligent aliens might be correct. It's just that the available evidence does not measure up to the standards of genuine science.)

Rather than hordes of aliens shouting over the radio or signaling with modulated laser beams from every potentially habitable planet, we are instead confronted with what scientists who study the possibility of extraterrestrial intelligence call "The Great Silence"—a confounding lack of any serious evidence that extraterrestrial life and intelligence actually exist anywhere in the universe except on Earth.

But let's step back a moment and ask whether this absence of evidence should really surprise us. SETI scientists argue that the serious search for extraterrestrial intelligence has only barely begun, and that billions of stars in our own Milky Way galaxy have yet to be carefully scrutinized for telltale traces of a technologically advanced civilization. Beyond the Milky Way are billions of other galaxies floating in the fathomless ocean of interstellar space. Surely it is too soon to conclude that all are bereft of the spark of life and mind. After all, as SETI scientists are fond of telling each other in a kind of self-reassuring mantra, "Absence of evidence of ETI is not evidence of absence of ETI." Maybe after a few more millennia of deafening interstellar silence it might be fair to conclude that mankind is probably alone in the uncaring vastness of the universe, but such a depressing verdict is surely premature.

A Market-Based Approach to Narrowing the Search for ETI

In *The Wisdom of Crowds*, James Surowieki recounts the novel approach used by a naval officer named John Craven to locate the final resting place of a missing American submarine, the USS *Scorpion*:

> First, Craven concocted a series of scenarios—alternative explanations for what might have happened to the *Scorpion*. Then he assembled a team of men with a wide range of knowledge, including mathematicians, submarine specialists, and salvage men. Instead of asking them to consult with each other to come up with an answer, he asked each of them to offer his best guess about how likely each of the scenarios was.... And so Craven's men bet on why the submarine ran into trouble, on its speed as it headed to the ocean bottom, on the steepness of its descent, and so forth.[6]

As Surowieki notes, while none of these factors would definitively indicate the location of the lost sub, their summed totality, if reasonably accurate, might offer a valuable clue:

> Needless to say, no one of these pieces of information could tell Craven where the *Scorpion* was. But Craven believed that it he put all the answers together, building a composite picture of how the *Scorpion* died, he'd end up with a pretty good idea of where it was.[7]

The group's collective estimate of where the lost submarine had come to rest turned out to be uncannily accurate: The wreck of the *Scorpion* was found 220 yards from where Craven's ensemble of betting men predicted it would be. Although no single individual in the group had selected the favored location, the crowd in its collective wisdom had focused, laser-like, on the right spot on the ocean floor. Miraculously, it had done so on the basis of very little data:

> What's astonishing about this story is that the evidence that the group was relying on in this case amounted to almost nothing.... No one knew why the submarine sank, no one had any idea how fast it was traveling or how steeply it fell to the ocean floor. And yet even though no one in the group knew any of these things, the group as a whole knew them all.[8]

The eerie success of John Craven's unorthodox approach is scarcely the only evidence that a suitably structured and properly motivated crowd will inevitably display a level of group intelligence that vastly exceeds the wisdom of even the smartest individuals within the group. As *The Wisdom of Crowds* documents in exhaustive detail, the phenomenon of the emergence of what is essentially superhuman group intelligence is undeniable.

One scientific challenge that could benefit from the application of superhuman group intelligence is the daunting task of narrowing the search for extraterrestrial intelligence (SETI) to a manageable list of target stars and interception modalities. As with the enormous initial search space that confronted the seekers of the lost *Scorpion*, the cosmic haystack facing SETI researchers is stupendously large. As Jill Tarter wrote recently in an article in *Skeptical Inquirer*:

> The Cosmic Haystack is large, unimaginably large, and is at least nine-dimensional. And that's only the haystack we can describe today with what we know about physics and technology in the twenty-first century, and from our terrestrial and inescapably anthropocentric vantage point. The fact that we've so far pulled a few straws from that haystack, examined them, and declared that no "needle" has yet been found doesn't make the haystack any smaller, nor invalidate the reasons we set out to try to explore it in the first place.[9]

The appropriate response to this challenge on the part of the SETI community is the development of better search tools, Tarter believes:

> If the history of the Search for Extraterrestrial Intelligence (SETI) argues for anything, it argues for better search tools....[10]
>
> The set of "better search tools" might profitably include not only more powerful radio-telescopes dedicated to SETI, such as the new Allen Telescope Array (*www.seti.org/ata*) and a new optical SETI sky survey instrument at Harvard (*http://seti.harvard.edu/oseti*), but also market-based techniques for narrowing the search space, analogous to the technique used by John Craven to locate the missing *Scorpion*. And who better to enlist as market participants than the globally diverse, fiercely free-thinking group of more than 3 million volunteers who participate in the SETI@Home project?

Then again, there is another exotic possibility: Evidence of alien intelligence might be in plain view but completely unrecognizable to us. Is this suggestion plausible? I submit that it is not merely possible, but virtually certain that we would fail to recognize evidence of the existence of a sufficiently advanced alien intelligence, even if it were right under our noses.

In a recent paper, Spanish theoretical physicist Beatriz Gato-Rivera suggested a startling possibility that would explain the Fermi Paradox in a radically novel but entirely rational way: The human civilization of planet Earth might possibly be immersed in a much larger and far more highly evolved civilization of which we are blissfully unaware.[11] A variant on the so-called "cosmic zoo" hypothesis—the notion that Earth is a protected enclave deliberately shielded from premature contact with more advanced civilizations by an intergalactic version of the Sierra Club—Gato-Rivera's idea is that Earth could be the cosmic counterpart of an anthill or a prairie dog community located alongside an interstate freeway—tolerated, maybe even protected by solicitous extraterrestrial environmentalists, but more likely ignored (except, perhaps, as an object of occasional study by biologists specializing in the primitive exotica to be found in luxuriant abundance among earthly flora and fauna).

As planetary scientist David Grinspoon puts it in his engrossing book *Lonely Planets: The Natural Philosophy of Alien Life*:

> Maybe an advanced civilization long ago spread throughout the galaxy, but to them we are so clearly not intelligent, and incapable of meaningful conversation, that they don't bother with us. To the truly intelligent species in the galaxy, we may not seem threatening or promising.[12]

Would we necessarily even know if our "civilization" were embedded in a stupendously advanced technological supercivilization? After all, as Gato-Rivera points out, mountain gorillas surely do not realize that their "civilization" is embedded in a 21st-century world of Internet communication, transcontinental jet travel, and interplanetary robotic exploration.

The Possibility of Undetectable Patterns in Astrophysical Data

One intriguing possibility is that markers of extraterrestrial intelligence are buried undetectably in the mountains of astrophysical data that already exist. These data collections, which are growing exponentially in both volume and complexity, may contain evidence of artificially generated patterns that simply exceed our limited human capacity to detect or decipher them.

Recognition of human limits on the ability to comprehend such complex data sets is one of the key motivations behind the Virtual Observatory project, which would enlist novel artificial intelligence tools to assist human beings in making sense of this deluge of new information about the universe. As Caltech scientists S. G. Djorgovski and R. Williams wrote in a recent scientific paper titled "Virtual Observatory: From Concept to Implementation":

> The Virtual Observatory (VO) concept is the astronomical community's response to [the challenge of exponentially increasing amounts of astrophysical data]. VO is an emerging, open, web-based, distributed research environment for astronomy with massive and complex data sets. It assembles data archives and services, as well as data exploration and analysis tools. It is technology-enabled, but science-driven, providing excellent opportunities for collaboration between astronomers and computer scien[tists].... It is also an example of a new kind of a scientific organization, which is inherently distributed, inherently multidisciplinary, with an unusually broad spectrum of contributors and users.[13]

The Virtual Observatory contributors and users, Djorgovski and Williams assert, will necessarily include some form of artificial intelligence:

Most [astrophysical] data and data constructs, and patterns in them, cannot be comprehended by humans directly. That is a direct consequence of a growth in complexity of information, mainly its multidimensionality. This requires the use or development of novel data mining or knowledge discovery in databases and data understanding technologies, hyper-dimensional visualization, etc. The use of AI/machine-assisted discovery may become a standard scientific practice.[14]

The two scientists believe that using artificial intelligence to overcome human biological limitations on the visualization of complex data patterns may lead to some crucial new astrophysical insights. It seems plausible that one such insight might be the detection of hyper-dimensional patterns of astrophysical data marking the presence of advanced intelligence—patterns that may be invisible to our biologically limited eyes and minds. If so, the Virtual Observatory could turn out to be one of the most important SETI research tool ever envisioned.

In answer to the question of why a hypothetical advanced extraterrestrial civilization has not seen fit to make contact and enlighten us about the secrets of the universe, Gato-Rivera offers a disconcerting but eminently sensible response:

> The answer to the usual remark "if there are advanced extraterrestrials around, why do they not contact us openly and officially and teach us their science and technology?" seems obvious. Would any country on this planet send an official delegation to mountain gorilla territory to introduce themselves "openly and officially" to the gorilla authorities? Would they shake hands, make agreements and exchange signatures with the dominant males? About teaching us their science and technology, who would volunteer to teach physics, mathematics and engineering to a bunch of gorillas?...How many bananas would be necessary for the most intelligent gorillas to understand the Maxwell equations of electromagnetism?...In the same way we may wonder how many sandwiches, potato chips or cigarettes would be necessary for the most intelligent among our scientists to understand the key scientific and technological results of a much more advanced civilization. Our intellectual faculties...are limited

> by our brain capabilities that are by no means infinite. Therefore it is most natural and sensible to assume that there may exist important key scientific and technological concepts and results whose understanding is completely beyond the brain capabilities of our species, but is within reach of much more evolved and sophisticated brains corresponding to much more advanced civilizations.[15]

Instead of eagerly revealing themselves to earthly observers (including SETI researchers), advanced extraterrestrial civilizations might instead opt for deliberate concealment and avoidance of contact (just as we put up window screens and mosquito nets to discourage "contact" by insect pests). Indeed, Gato-Rivera hypothesizes, advanced alien civilizations might have decided long ago to deliberately shield themselves from the prying eyes of SETI researchers located on Earth and elsewhere, thus offering another plausible explanation for the Fermi Paradox:

> The Undetectability Conjecture. Generically, all advanced enough civilizations camouflage their planets for security reasons, so that no signal of civilization can be detected by external observers, who would only obtain distorted data for dissuasion purposes.[16]

In that case, the absence of evidence of the existence of intelligent aliens that troubled Fermi would simply be a predictable consequence of the fact that our primitive observational tools are being thwarted by sophisticated alien cloaking technology. As Gato-Rivera puts it:

> The right claim would be in this case that there is no signal of *primitive civilizations*, like ours, who would allow themselves to be detected by external observers, but *nothing* can be said about the possibility of *advanced civilizations*, capable of fooling our telescopes, detectors and space probes, and who would not allow themselves to be detected.[17]

Adding to the plausibility of Gato-Rivera's Undetectability Conjecture is a recent analysis by NASA chief historian Steven J. Dick, published in the *International Journal of Astrobiology*, which concludes that if extraterrestrial intelligence exists, it has very probably evolved beyond flesh and blood intelligence to an advanced form of artificial intelligence that is the product of an extremely long process of technological and cultural evolution. In such a post-biological universe, machines would be the predominant form of intelligence; intelligent biological beings immersed in technological civilizations capable of interstellar signaling would be a minuscule minority:

> Biologically based technological civilization...is a fleeting phenomenon limited to a few thousand years, and exists in the universe in the proportion of one thousand to one billion, so that only one in a million civilizations are biological.[18]

What we are really searching for in the SETI project, it turns out, is something far rare than mere extraterrestrial intelligence. What we are actually seeking is alien intelligence that has reached roughly our current level of development—alien beings who are not too dumb and not too smart in comparison to humanity to be recognizable as fellow sentient creatures—and that evolved through a process similar enough to terrestrial evolution that humankind and the putative ETs will have some rudimentary common frame of reference.

What are the chances that this drastically narrowed search pattern will succeed? Rather slim, I suggest. Perhaps a twin Earth exists somewhere out there, but the odds that we will stumble across it seem vanishingly small.

Genocidal ETs: The Doomsday Scenario

It must be mentioned that there is one horrific scenario on offer from the SETI research community that explains "The Great Silence" in a way that is all too familiar to historians who study the depressing record of human-induced species extinction and genocide on Earth. What if our vastly superior cosmic neighbors are none too neighborly toward lesser races deemed to be potential pests? What if, instead of merely controlling the activities of human beings and other living vectors of potentially dangerous infestations (terraforming, nanotechnology, and nuclear pollution among them) with the cosmic equivalent of humane pest control methods, they instead prefer to use something more lethal and permanent to eliminate the risk of contamination? Perhaps something functionally akin to planetary DDT.

On Earth, we are confronted daily with unimpeachable evidence that, with rare exceptions, human beings will not flinch at snuffing out the life of a fly or a beetle, or even a large mammal. Why should our unmitigated disdain for the rights of lower orders of life not be shared by beings whose intellects are to human minds as our intellects are to the mind of an insect or a mollusk?

The good news is that there is no sense loosing sleep over this doomsday scenario. If we are slated for extermination at the hands of a higher order of intelligent creatures, there's probably not much we can do about it. And if such implacable evil exists in the universe, our best bet is to progress technologically as rapidly as possible in order to be able to defend ourselves against it.

Yet we should not be unduly discouraged about the chances of someday resolving the Fermi Paradox and punctuating "The Great Silence" with genuine interstellar dialogue. I submit that if living biospheres are reasonably common throughout the cosmos, there is reason to hope that the SETI project will eventually succeed, although probably not in our lifetimes. That hope is based on three very recent theories about the nature of computation and evolution (which can be characterized as an extremely complex form of computation): (1) MIT physicist Seth Lloyd's conclusion that there exists a theoretical upper limit on the amount of computation that can occur in our universe, given the prevailing laws of physics and such constraining physical factors as the speed of light and the gravitational constant;[19] (2) Futurist Ray Kurzweil's conjecture that intelligent computers will be propelled into an exponentially quickening process of artificial evolution by what he calls the Law of Time and Chaos until they roar past the human level of intelligence and soar far beyond our biologically constrained intellectual capacities;[20] (3) Cambridge University theoretical biologist Simon Conway Morris's hypothesis that evolutionary forces in disconnected ecosystems invariably converge on similar solutions to the common challenges confronted by living creatures—the emergence of eyes (or organs roughly analogous to eyes) with which to perceive electromagnetic radiation and use its patterns of fluctuation to probe the changing shape of the physical world, ears with which to register and map the sources of sonic vibrations, wings with which to navigate a gaseous or aqueous atmosphere, noses with which to sample chemical signals in the environment, and appendages with which to propel the organism through that environment and seize chunks of it for nourishment or shelter.[21]

If all three hypotheses are correct, and if processes of biological and technological evolution have taken hold on hospitable planets throughout the universe, then I suggest that it is almost a foregone conclusion that some form of *cosmic cultural convergence* will eventually manifest itself.

What do I mean by cosmic cultural convergence? As the universe continues to age and as evolution progresses in hypothetical biospheres scatter throughout the cosmos, sentient beings (biological or mechanical) will, in accord with Kurzweil's Law of Time and Chaos, continue to get smarter and smarter at an exponentially quickening pace. Yet they will never become *infinitely* intelligent because, as Seth Lloyd's analysis demonstrates conclusively, the basic laws and constants of physics impose an absolute upper limit on the amount and speed of computation that can ever occur in our universe. Sentient organisms throughout the universe (ourselves and our probable robotic progeny included) will eventually hit an insurmountable intellectual wall that will place unavoidable maximum limits on their ability to compute. Thus, even though some extraterrestrial civilizations may have become math and physics whizzes many millennia before proto-humans dropped out of the trees, all the civilizations of the cosmos (or at least those that survive to a reasonable maturity) will eventually max out at a roughly equivalent level of intellectual attainment. That seems to be the logical implication of Lloyd's conclusions about the theoretical upper limits to computation.

Finally, as Simon Conway Morris has demonstrated, because evolution tends to independently reinvent common solutions to common problems over and over again, there is reason to believe that cultural evolutionary trends among civilizations scattered throughout the cosmos will ultimately converge in a manner that will make their cogitations and communications mutually comprehensible. This seems especially likely for those planetary civilizations destined to venture out into the frontier of outer space where they will confront identical challenges: development of interstellar propulsion technology, perfection of navigational techniques, improvements in the accuracy of astronomical observation, and mastery of other technologies necessary to accomplish long-distance space travel.

The Laws and Constants of Physics as a Cosmic Rosetta Stone

At a 2004 SETI conference convened in Atlanta (under the auspices of the American Anthropological Association), scientists representing a wide range of disciplines discussed the topic "Anthropology, Archaeology, and Interstellar Communication: Science and the Knowledge of Distant Worlds."[22] SETI Institute social scientist Douglas Vakoch reported the sobering but unsurprising conclusion of the conferees: "It may be much more difficult to understand extraterrestrials than many scientists have thought before."[23] The principal reason for skepticism was the difficulty historians experienced decoding Egyptian and Mayan hieroglyphics. As anthropologist Ben Finney of the University of Hawaii noted, the task of decoding these ancient writings was delayed because of erroneous baseline assumptions about communication protocols—for many years, the individual symbols were regarded as denoting an entire idea or concept rather than elements in a language—and was finally accomplished only with the aid of such keys as the multilingual Rosetta Stone.

Many SETI scientists are more sanguine than Vakoch, believing that the laws and dimensionless constants of physics—which seem, despite some recent speculations to the contrary, to be invariant throughout the visible universe—might provide a kind of cosmic Rosetta Stone that would allow abstract symbolic communication between civilizations with nothing more in common than their capacity to probe the natural environment by means of scientific investigation.

Put differently, there is at least a plausible hope that extraterrestrial civilizations and our own terrestrial civilization will eventually evolve toward a roughly equivalent state of intellectual competence, and that the forces of cultural evolution will someday, if only in the far distant future, converge in a manner that will make genuine interstellar communication possible, even among species that began the long trek toward sentience at very different starting points in time and space. If it eventually occurs, this moment of convergence might conceivably prove to be the opening motif in a cosmic concert of cultures—the sounding of a deep chord heralding the birth of a cosmic community.

But as for the prospect of meaningful communication with ET in the short-term—in an era when we and other intelligent creatures that may exist throughout the universe are, in all likelihood, mere infants babbling away in mutual incomprehensibility on our widely scattered and evolutionarily disconnected "lonely planets" (to borrow David Grinspoon's evocative phrase)—I remain skeptical.

Chapter 5

COLONIZING THE COSMOS AT LIGHT SPEED

Let's put aside for a moment the rather discouraging conclusion of the previous chapter and assume for the sake of argument that some ET, somewhere in the vastness of the universe, does decide that it would like to reach out and touch other intelligent species that might reside in this galaxy or the next. If a highly evolved extraterrestrial civilization were to decide to communicate with its potential peers around the universe, how might it do so?

The conventional wisdom among SETI researchers is that ET would send radio or optical signals, then wait patiently for the encoded photons to travel decades or centuries across interstellar space in the faint hope that someone or something, on some random planet in this galaxy or another, will eventually pick up the phone and reply with an intelligible answer. More boring decades or centuries would then ensue while the answer wends its way back to the transmitting planet. And an equally insufferable interval of silence would transpire before the initiator could respond to the answer.

This scenario, which might be called "The Long Hello," would surely tax the patience of even the most ardent alien conversationalist.

Artificial Exo-Society Modeling

If "The Long Hello" scenario does come to pass, then it will be essential for SETI scientists to compensate for the impact of cultural evolutionary forces that will likely play out in ET societies during the years, decades, or even centuries separating message exchanges. We will surely want to figure out the direction in which ET's culture is evolving before we entrust the rascals with our deepest secrets.

In a paper I delivered to the SETI II Session of the 2002 International Astronautical Congress in Houston,[1] I suggested that one way to approach this challenge would be to bring to bear the tools of the newest field of complexity research: artificial society modeling. Methodologically related to artificial life research, artificial society modeling utilizes agent-based computer simulation tools such as SWARM and SUGARSCAPE (developed by the Santa Fe Institute, Los Alamos National Laboratory, and the Bookings Institution) in an effort to introduce an unprecedented degree of rigor and quantitative sophistication into social science research. The broad aim of artificial society modeling is to begin the development of a more unified social science that embeds cultural evolutionary processes in a computational environment that simulates demographics, the transmission of culture, conflict, economics, disease, the emergence of groups, and coadaptation with an environment in a bottom-up fashion. When an artificial society computer model is run, artificial societal patterns emerge from the interaction of autonomous software agents (the inhabitants of the artificial society).

Joshua Epstein, a leading artificial society modeling expert affiliated with the Brookings Institution and the Santa Fe Institute, made this observation about the close linkage between computation and agent-based artificial society modeling:

> The agent-based approach invites the interpretation of society as a distributed computational device, and in turn the interpretation of social dynamics as a type of computation.[2]

Artificial society modeling techniques offer the potential of computational simulation of hypothetical alien societies in much the same way that artificial life modeling techniques offer the potential to model hypothetical exo-biological phenomena. NASA recently announced its intention to begin exploring the possibility of including artificial life research within the broad portfolio of scientific fields encompassed within the interdisciplinary astrobiology research endeavor. It may be appropriate for SETI researchers to likewise commence an exploration of the possible inclusion of artificial exo-society modeling within the SETI research endeavor.

Artificial exo-society modeling might be particularly useful in a post-detection environment by: (1) coherently organizing the set of data points derived from a detected ETI signal, (2) mapping trends in the data points over time (assuming receipt of an extended ETI signal), and (3) projecting such trends forward to derive alternative cultural evolutionary scenarios for the exo-society under analysis. The latter exercise might be particularly useful to compensate for the inevitable time lag between generation of an ETI signal and receipt of an ETI signal on Earth.

But there are more exotic possibilities that would avoid the decades or centuries of punctuating silence. For instance, the distinguished Australian astrophysicist Paul Davies has suggested that an enterprising extraterrestrial biologist might have hit upon the ingenious strategy of embedding a message in the DNA of extraterrestrial viruses, then dispatched them to Earth in tiny spaceships (or even meteorites) with the intent of infecting the DNA of earthly organisms. If the alien viral DNA took root in the genes of earthlings, it could be unwittingly passed down through thousands or millions of generations until terrestrial creatures finally evolved sufficient intelligence to decode the message. The content of the message, which could conceivably lurk undetected for millennia in highly conserved stretches of so-called "junk" DNA, might be contained in its entirety within the viral DNA segment. Alternatively, the segment might be a kind of map pointing to the location, in time or space, of a larger message secreted elsewhere—perhaps in a highly compressed data-rich radio or optical signal broadcast at precisely described intervals each century or two, or even hidden away in a physical cache located "safely in the fringes of our solar system, in which we would find the entire contents of an Encyclopedia Galactica, including the rise and fall of ET's civilization, which may have died out long before human beings even existed."[3]

This approach would seem to limit ET to a one-way message. Or would it? If either the embedded DNA message or a larger message referenced by the viral signal constituted an instruction set for *growing* something—particularly something that could evolve and learn—then some semblance of a dialogue, up close and personal, might eventually be possible with whomever or whatever eventually emerged from a decoding of the basic message.

Davies may believe that his proposal is at the extreme end of the spectrum of speculation but even wilder ideas have been entertained by mainstream researchers about how to profitably search for ET's message in non-obvious venues. Perhaps the most ingenious effort is led by Canadian SETI researcher Allen Tough who believes that some form of extraterrestrial intelligence may already be monitoring (or perhaps even inhabiting) the untamed virtual frontier of the World Wide Web. Tough theorizes that just as the National Security Agency is capable of monitoring fax, email, and telephone messages around the world, an advanced intelligence of extraterrestrial origin "will have little difficulty learning our languages and surfing the Web as competently as we do. If it uses the major search engines to find web pages on *extraterrestrial intelligence, alien intelligence, alien probe,* or *invitation to ETI,* for instance, it will find any invitations [to communicate with earthlings] that exist on the Web."[4]

Proceeding on this assumption, Tough and a group of his SETI colleagues have actually posted a warm welcome to ETI and an invitation to engage in dialogue with humankind on the Web at *http://ieti.org/hello.html.* Here are some noteworthy excerpts from their invitation:

> Greetings to extraterrestrial intelligence!
>
> If you originated in some other place in the universe, we welcome you here. And we invite you to establish communication with us and with all of humanity. We enthusiastically look forward to that dialogue.
>
> We hope you will dialogue with humankind about science and society and philosophy. About the universe. About your culture and ours. About the biggest questions of all. In addition, we welcome your advice on how we could successfully switch to wise ethical ways of dealing with conflicts, our natural environment, the well-being of future generations, and other global issues and opportunities.
>
> We will treat you with respect, courtesy, friendship, and caring. We will speak and act truthfully, avoiding lies and deception. We will deal honestly and fairly with you, avoiding any temptation to exploit the situation for personal greed or for any particular nation or organization. Without forsaking our own values and integrity, we will be as empathic, helpful, and flexible as we can in understanding and fostering your goals and plans.
>
> You may have some major reason for avoiding a dialogue with humanity at this stage. Maybe you do not want to unduly influence the natural course of our cultural development, for example. Or maybe you have little interest in most aspects of our culture. Or maybe you perceive our preparations for contact and interaction as inadequate. In any case, we hope you will tell us your reasons. It might even be possible, working together, to find some creative solution or compromise. We are eager to do whatever will enable successful interaction.

So far, ET hasn't hit the reply button.

Searching for ET in the Computational Universe

Computer scientist Stephen Wolfram believes that traditional SETI researchers are on the wrong track when they scan the skies looking for radio or optical signals that would betray the presence of advanced extraterrestrial civilizations. The kind of evidence for which SETI scientists are searching—patterned signals analogous to the amplitude and frequency modulations used by AM and FM radio stations to encode audio information—are notoriously inefficient, and hence unlikely to be employed as methods of interstellar communication by technologically adept aliens, Wolfram believes. Indeed, as he noted in a 2005 *New Scientist* interview, the communications of an advanced alien civilization are likely to closely resemble the random noise we associate with astrophysical sources such as stars and interstellar gas clouds.[5] In this respect, they will have the random appearance of terrestrial signals that have been digitally compressed using technologies such as code division multiple access (CDMA), a widely employed cell phone compression protocol. Accordingly, it will be extremely difficult (and perhaps impossible) to distinguish the transmissions of a highly advanced alien civilization from random cosmic noise generated by natural sources.

Wolfram thinks that a better SETI strategy is to begin searching what he calls the computational universe for evidence of ET's existence. The computational universe is an enormous imaginary library containing all the computer algorithms that could possibly exist, together with all the outcomes that could possibly be produced by running those algorithms for a very long time (at least as long as the age of the universe). Within that computational universe, Wolfram believes, is to be found not only the human species together with all of its cultural artifacts, but also every alien species and alien civilization that could possibly exist.

This notion—a rather fanciful extrapolation from Wolfram's premise that a kind of Software of Everything is responsible for generating all the complexity and order that we observe in the cosmos—would seem to require the mother of all mainframes to handle the calculations necessary to flesh in all of the complexities of the computational universe that Wolfram has imagined. I am skeptical that such a computer could actually be constructed and, even if constructed, that it could perform the requisite computations any faster than the vast natural quantum computer that is the cosmos is currently performing those computations. So my gratuitous and unsolicited advice to the SETI research community is this: Disregard Wolfram's discouraging words, keep scanning the skies, but do not categorically ignore signals that may appear to be little more than random noise. Those quasi-random signals may turn out to be the distinctive signature of a truly advanced alien civilization.

Finally, in a cover story published in the August 25, 2004, issue of the prestigious scientific journal *Nature*, Rutgers University researchers Christopher Rose and Gregory Wright predicted that ET was likely to eschew electronic modes of communication altogether and use some form of snail-mail. Why? Because the expense per bit of transmitted information would supposedly be less in terms of the power expended if a message of massive length were sent by (relatively) slow interstellar rocket rather than via a light-speed radio or optical transmission. But as SETI scientist Seth Shostak has pointed out, there are severe offsetting problems and expenses that would plague the snail-mail approach to interstellar communication. For one thing, the "postal rocket" would have a sizeable cost (as in trillions of dollars), offsetting any theoretical price advantage that physical transmission might enjoy over electromagnetic beaming. Second, during its 100,000-year travel time to a stellar system of interest, the target star and its planets will likely have been nudged into unpredictable locations by subtle gravitational interactions within the star's own solar system. How would the postal rocket be able to home in on the correct destination address once it arrives in the stellar neighborhood? All in all, Shostak concludes, radio or light-wave communication makes more sense than snail mail: "Sometimes it's better to eschew the Pony Express, and saunter down to the telegraph office."[6]

Assuming ET's message isn't already here—lurking on the Internet, secreted in the vast stretches of junk DNA inside our cells, or hidden in the crevices of a distant asteroid orbiting at the edge of our solar system—and further assuming that snail-mail communication with a distant civilization would impose insurmountable objects (boredom paramount among them), what strategies might a highly evolved alien intelligence employ to both communicate with creatures such as ourselves, and avoid the problem of intolerable delay between episodes of communication?

The first option might be called interstellar cloning. In a forthcoming scientific paper, I will suggest that a truly advanced extraterrestrial civilization might not transmit a "message" in the conventional sense, but rather a sophisticated software program constituting a detailed instruction set for replicating individual members of that civilization or, better yet, the entire civilization and its supporting biosphere. Such an instruction set would essentially constitute the genome—cultural as well as biological—of the transmitting civilization. This possibility it not as outlandish as it might initially appear.

Nobel laureate James Watson pointed out in his memoir, *DNA: The Secret of Life*, that *life* is simply the emergent property of suitably encoded matter, animated by thermodynamic disequilibrium. If life's code can be reliably transmitted across *time* (by means of a succession of generations of organisms that serve as high-fidelity DNA duplication devices), then why could it not also be reliably transmitted across *space* (by means of radio or optical transmission of a data set constituting replication instructions for a particular organism or even an entire biosphere)?

If the transmission were sufficiently robust and precise, it would amount to a kind of interstellar cloning process: a means by which an advanced society could spread its cultural and biological DNA across the universe at light speed and without the risk of physical travel. In a very real sense, this kind of message from ET would be ET itself (or at least a recipe for reconstituting ET from raw materials indigenous to the environment of the receiving civilization).

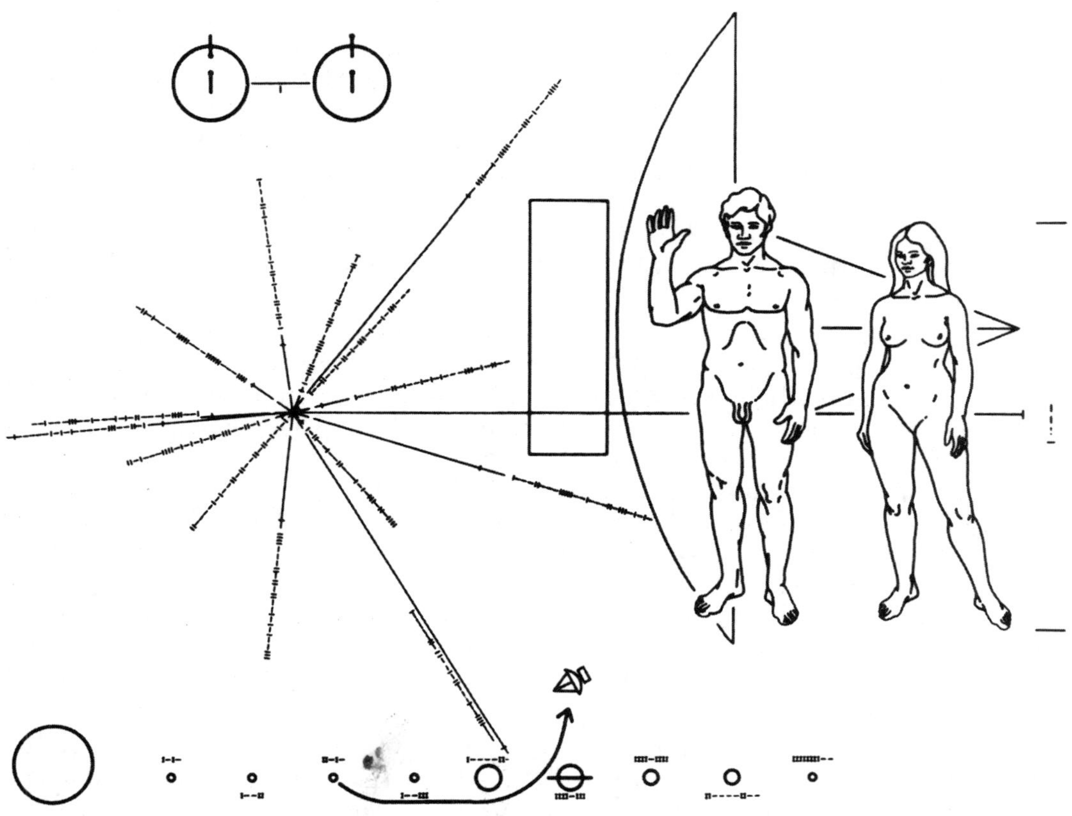

When NASA's Pioneer 10 spacecraft was launched in 1977, it carried a 12-inch disk bearing words, music, and images (selected by a team of scientists) to represent Earth.
Image provided by Wikipedia

The interstellar cloning approach has three distinct advantages. First, only a single transmission would be required. Following receipt, ET in all its glory (perhaps accompanied by an entire ecosystem and surrounding civilization) could, at least in principle, be completely reconstituted by the recipients of the message. Alternatively, ET and its surroundings could be brought to artificial life within the confines of a sufficiently commodious computer simulation system. Genuine dialogue could then commence instantaneously, albeit with the version of ET and the transmitting civilization that existed on the date of transmission.

Second, if we assume that a prime objective of the transmitting civilization is to spread its genes and memes throughout the vastness of interstellar space by means of a process that might be called *directed virtual panspermia*, then a message consisting of the genetic and cultural DNA of the transmitting civilization would be a vastly superior means of propagation in comparison to the broadcast of a low-bandwidth message such as "Hello! Anybody out there?"

Third, the interstellar cloning approach would virtually guarantee that the reproductive seeds of the transmitters would fall on fertile soil. Why? Because only a sufficiently advanced civilization would have the technological wherewithal to even recognize the transmission as an artificial message, let alone the capacity to decipher it and then reconstitute the transmitting ET and his ecosystem. And only an advanced civilization that was sufficiently curious would bother to attempt to solve the puzzle posed by receipt of such a message in the first place. These very qualities—advanced technological capacity and insatiable curiosity—would seem to be the essential attributes of a receiving civilization likely to serve as hospitable potting soil for alien seeds. In short, the very nature of the transmission would tend to ensure that the virtual DNA of the alien civilization would either take root or be disregarded entirely.

Several science fiction writers have explored this scenario and identified ominous possibilities. In 1962, cosmologist Fred Hoyle coauthored a novel titled *A for Andromeda*, in which British radio astronomers stumble upon an intelligent transmission emanating from the Andromeda galaxy. The message turns out to be a three-part instruction manual for first building an advanced computer, then installing an operating system, and finally downloading what amounts to application software capable of sophisticated interactions with earthlings. The extraterrestrial software is a kind of genetic algorithm, endowed with the capacity to learn and evolve at breakneck speed. Seemingly benign at first, the vastly superior alien intelligence whose emergence is programmed into the intercepted extraterrestrial message turns out to have nefarious intentions—to supplant human control of planet Earth, thereby assuring the proliferation and supremacy of its form of life and mind.

Something similar to the Hoyle scenario seems eminently plausible if we are dealing with a form of intelligence that has evolved in such a way as to maximize its capacity to replicate and dominate its environment. Indeed, the receipt of any extraterrestrial message that appears to be an instruction set for building or growing anything whatsoever, such as a potentially lethal microbe or a disabling computer virus of unfathomable power, should be treated with great caution. Credulous SETI researchers may *want* to believe that ET's intentions will be benign and altruistic—that any advanced alien intelligence we are likely to encounter will turn out to be a kind of secular deity from whom we can learn the secrets of world peace and the hidden meaning of the universe. They may fondly hope that a real-life version of *Star Trek*'s prime directive of non-interference will constrain the aliens' baser motives. However, the unambiguous record of evolutionary history on our own planet teaches a contrary lesson. The essence of that lesson is that natural selection is an implacable force that shapes us all—bacteria, plants, humans, and the remainder of the living firmament—into relentlessly motivated survival machines. Why should the same principle not hold sway on a cosmic scale? As evolutionary theorist Richard Dawkins has contended, Darwinian evolution through natural selection may not be merely incidentally true of our kind of life, but is almost certainly true of all life, everywhere in the universe. Natural selection–driven evolution may well turn out to be a cosmos-wide phenomenon, and Darwinian evolution may really matter in the universe, according to Dawkins. The inescapable implication is that ET, like humankind, may well have been molded by the forces of natural selection into a lean, mean, replication and survival machine that would not flinch at the prospect of either domesticating or eliminating competitor species.

Undaunted by this disquieting possibility, SETI Institute researcher Douglas Vakoch has launched a project funded by the John Templeton Foundation to draft possible responses to an intercepted message from ET. Vakoch's draft messages focus on "using the language of math and science to compose interstellar messages that describe altruism as we encounter it on Earth."[7] The idea, more or less, is to discover ways to put humanity's best foot forward in an opening volley of exchanged messages—figuring out how to use the language of mathematics, for instance, to sketch out some of our more appealing human qualities "like our ability to love"—while suppressing references to our less lovable traits (for example, our depressingly well-documented proclivity to employ the tools delivered to us by advanced technology to wage ever more lethal genocidal warfare).

Can we reasonably anticipate that a highly evolved form of extraterrestrial intelligence will be more candid about the dark side of its nature than we are prepared to be? Isn't it perfectly reasonable to anticipate that any rational ET we are likely to encounter will gloss over the scarier aspects of its nature and emphasize the positive—posturing itself exactly as Vakoch urges us to portray humanity in an interstellar message: benignly motivated by "our aspirations to be more altruistic—both to those close to home and to those more distant from us."[8]

The ET Google Strategy: Let Them Eat the WWW

A radically different communication strategy was recently advocated by SETI Institute scientist Seth Shostak. Rather than a carefully manicured and meticulously composed message purged of references to the darker aspects of human nature, Shostak suggested that we simply feed any potential alien listeners the unfiltered content of the entire World Wide Web:

> So here's my take on message construction: Forget about sending mathematical relationships, the value of pi, or the Fibonacci series. Rid your brain of the thought (no doubt borrowed from "Close Encounters of the Third Kind") that aliens are best addressed with musical arpeggios. No, if we want to broadcast a message from Earth, I propose that we just feed the Google servers into the transmitter.

> Send the aliens the World Wide Web. It would take half a year or less to transmit this in the microwave; using infrared lasers shortens the broadcast time to no more than two days.[9]

Shostak concedes that this approach would expose our frailties as well as our strengths, but is untroubled by the prospect:

> Sure the Web contains a lot of redundant information (and a lot of unsavory material, too, but after all, that's part of the human condition).[10]

The skeptical lawyer in me warns that this would be false advertising in the extreme—a seriously misleading caricature that distorts the complexities of the human spirit—aimed at intentionally deceiving the recipients of the message. My inner conspiracy theorist whispers that an alien civilization could play at this deception game with at least equal skill—and might well be motivated to do so, for reasons either benign (perhaps to simply convey a favorable first impression) or malicious (perhaps to assuage suspicions or disarm safeguards impeding penetration of computer systems by alien software viruses).

In any event, the factor of human curiosity is likely to overwhelm every precautionary instinct. If we someday receive a coherent message of extraterrestrial origin, will we heed the warnings of pessimists and skeptics? Or, like Pandora, will we feel irresistibly compelled to lift the lid of the mysterious gift box from the stars and peek inside? My guess is that we will throw caution to the winds and rush headlong into the adventure of deciphering ET's message. Indeed, insatiable curiosity may be the key character trait relied upon by a putative alien intelligence that aims to commandeer our advanced technology in order to clone itself across the vast reaches of interstellar space. Our profound curiosity, which is perhaps our most noble human trait, may turn out to be the very quality that makes us suitable potting soil for nurturing the seeds of extraterrestrial intelligence.

Human curiosity may present grave risks, but it is also the foundation of the scientific enterprise. Curiosity fuels our quest to extend the human understanding of nature to the outer ramparts of the vast cosmos. And scientific curiosity provokes us to wonder whether schemes of communication even grander than those discussed in this chapter are conceivable. I am speaking now of communication, not across the daunting light-years of our universe, but rather communication *between* the multiple universes comprising a theoretical ensemble called a *multiverse* or *meta-universe*.

Chapter 6

THE COSMIC MEDIUM IS THE MESSAGE

There has been an explosion of scientific interest in recent years in the possibility that we inhabit not merely a solitary universe, but a *multiverse*—an infinite multiplicity of disconnected cosmic domains that are spawned by a process of eternal inflation. Think of this spawning process as multiple Big Bangs going off like firecrackers on the Fourth of July, except that the explosions continue to detonate for eternity. And think of the disconnected domains as parallel universes, forever inaccessible to human observation, but perhaps susceptible to a subtle kind of future-oriented engineering.

As science writer Sharon Begley put it in a *Wall Street Journal* essay speculating about whether the universe's basic traits are random or inevitable, "Although 'universe' has traditionally meant one-and-only-one, advances in cosmology suggest there may be multiple 'pocket universes,' each a child of its own Big Bang."[1]

Inflation and Eternal Inflation

The concept of cosmic inflation was developed by MIT cosmologist Alan Guth in 1981 to explain the perceived absence of cosmic monopoles—exotic fundamental particles predicted by mainstream cosmological theories but that have never been observed in nature. Guth's ingenious idea was that a short period of superluminal expansion (expansion faster than the speed of light) shortly after the initial moment of the Big Bang would have had the effect of scattering the predicted monopoles far and wide, so that we would have little chance of finding them in the cosmic haystack. Later it was discovered that Guth's ingenious idea could be used to address what seemed to be a daunting mystery: the very special and seemingly fine-tuned nature of the Big Bang that launched our particular universe. Specifically, inflation appeared to solve the *horizon problem* (the mystery of why the temperature of the early universe appears to have been extremely uniform, despite the fact that different parts of it could not have been in causal contact because of the limit on thermal communication imposed by the speed of light), the *smoothness problem* (the puzzling uniformity of matter distribution and space-time geometry at the largest scales in our universe), and the *flatness problem* (the remarkable fact that our cosmos is almost exactly flat). By positing an initial period of extremely rapid cosmic expansion, inflation seemed to make each of these problems disappear.

Later, Russian-born physicist Andrei Linde came up with a variation on Guth's proposal called *eternal inflation*. Linde's basic idea was that inflation and the Big Bang didn't just happen once, but that the process is eternal, yielding an infinite number of Big Bangs and disconnected universes. The assembly of universes created by all these Big Bangs is termed a *multiverse* or a *meta-universe*.

Despite its attractive features and its consistency with observational evidence coming in from state-of-the-art instruments such as the Wilkinson MAP probe, some skeptics remain unconvinced about whether an epoch of early cosmic inflation actually took place. British mathematician and physicist Roger Penrose, for instance, doubts the initial motivation behind the hypothesis of inflation. Why, Penrose asks, is it appropriate to resort to the inflation hypothesis on the basis of its perceived aesthetic appeal? Is this idea inherently more beautiful than the notion that the Big Bang was, in fact, fine-tuned with extraordinary precision? And is some mathematical physicist's geeky idea of beauty a legitimate basis for making judgments about the nature of nature? Finally, what is so beautiful about the arbitrary insertion into the physics of the early universe an inflationary scalar field—or an infinite number of such scalar fields in the case of eternal inflation—"unrelated to other known fields of physics and with very specific properties designed only for the purpose of making inflation work."[2]

As Begley notes, this weird, counterintuitive viewpoint is motivated by several queer and interesting theoretical insights as well as by a body of quite startling observations, one of which varies wildly from the predictions of quantum theory.

Together, these insights and observations raise the astonishing possibility that our universe is not a loner, but the inhabitant of a vast invisible neighborhood composed of brother-and-sister, mother-and-baby universes, all of which are endowed with distinctive physical laws and constants (such as the inverse square law of gravitational attraction and the speed of light).

The most significant scientific evidence in support of the multiverse theories that have been advanced in recent years is the observation that the universe we happen to inhabit is strangely and—at least on the basis of fundamental physical theory—accidentally bio-friendly. As Mario Livio and Martin Rees have observed with regard to the question discussed in the two preceding chapters (the possibility of extraterrestrial life and intelligence), the fact that we are even engaged in this discussion implies something profound about our concept of the basic nature of the cosmos:

> Does extraterrestrial intelligent life exist? The fact that we can even ask this question relies on an important truth: The properties of our universe have allowed complexity (of the type that characterizes humans) to emerge. Obviously, the biological details of humans and their emergence depend on contingent features of Earth and its history. However, some requirements would seem generic for any form of life: galaxies, stars, and (probably) planets had to form; nucleosynthesis in stars had to give rise to atoms such as carbon, oxygen, and iron; and these atoms had to be in a stable environment where they could combine to form the molecules of life.[3]

As Livio and Rees note, things need not have turned out so propitiously in the aftermath of the Big Bang that created, seemingly *ex nihilo*, our particular universe:

> We can imagine universes where the constants of physics and cosmology have different values. Many such "counterfactual" universes would not have allowed the chain of processes that could have led to any kind of advanced life. For instance, even a universe with the same physical laws and the same values of all physical constants but one—a cosmological constant Λ (the "pressure" of the physical vacuum) higher by more than an order of magnitude—would have expanded so fast that no galaxies could have formed. Other properties that appear to have been crucial for the emergence of complexity are (i) the presence of baryons (particles such as protons and neutrons); (ii) the fact that the universe is not infinitely smooth, allowing for the formation of structure (quantified as the amplitude of the fluctuations in the cosmic microwave background, *Q*); and (iii) a gravitational force that is weaker by a factor of nearly

> 10^{40} than the microphysical forces that act within atoms and molecules—were gravity not so weak, there would not be such a large differences between the atomic and the cosmic scales of mass, length, and time.[4]

As these two scientists point out, a key challenge for 21st-century physicists is to figure out "which of these dimensionless parameters such as Q and Λ are truly fundamental in the sense of being explicable within the framework of an ultimate, unified theory and which are merely accidental."[5] To decide, in short, which of the so-called laws of nature are truly and forever invariant and which are merely *by-laws* of our particular branch of the eternal, outrageously fecund, and infinitely variegated multiverse.

The Biggest of the Big Questions

Beyond that daunting challenge—which amounts to coming up with an answer to Einstein's famous question of whether God had any choice in designing the laws of physics—lies an even deeper question. This particular question is, in the view of Columbia physicist Brian Greene, the deepest question in all of science. Renowned cosmologist Paul Davies agrees, calling it the biggest of the Big Questions. And just what is this momentous question? Not the mystery of life's origin, though the profundity of that particular puzzle prompted Charles Darwin to remark that it was probably forever beyond the pale of human comprehension. A dog, Darwin commented famously, might as easily contemplate the mind of Newton. Not the inscrutable manner in which consciousness emerges from the interaction and interconnection of neurons in the human skull, though a cascade of Nobel prizes will undoubtedly reward the teams of neuroscientists who achieve progress in understanding this phenomenon. And not even the future course of biological and cultural evolution on planet Earth, though the great Darwinian river is surely carving a course that today's most visionary evolutionary theorist will have difficulty even imagining. No, the question is more profound, more fundamental, less tractable than any of these. It is this: *Why* is the universe life-friendly?

We have been taught since childhood that the universe is a horrifyingly hostile place. Violent black holes, planets and moons searing with unbearable heat or deep-frozen at temperatures that make Antarctica look tropical, and the vastness of interstellar space dooming us to perpetual physical isolation from our nearest starry neighbors—this is the depressing picture of the cosmos beyond Earth that dominates the popular imagination.

This vision is profoundly wrong at a fundamental level. As scientists are now beginning to realize, the truly amazing thing about our universe is just how strangely and improbably life-friendly or anthropic it is. As Cambridge evolutionary biologist Simon Conway Morris puts it in his book *Life's Solution*, "On a cosmic scale, it is now widely appreciated that even trivial differences in the starting conditions [of the cosmos] would lead to an unrecognizable and uninhabitable universe."[6]

Simply put, if the Big Bang had detonated with slightly greater force, the cosmos would be essentially empty by now. If the primordial explosion had propelled the initial payload of cosmic raw materials outward with slightly lesser force, the universe would long ago have recollapsed in a Big Crunch. In neither case would human beings or other life forms have had time to evolve.

As Stephen Hawking asks, "Why is the universe so close to the dividing line between collapsing again and expanding indefinitely? In order to be as close as we are now, the rate of expansion early on had to be chosen fantastically accurately."[7]

It is not only the rate of cosmic expansion that appears to have been selected, with phenomenal precision, in order to render our universe fit for carbon-based life and the emergence of intelligence. A multitude of other factors are fine-tuned with fantastic exactitude to a degree that renders the cosmos almost spookily bio-friendly. Some of the universe's life-friendly attributes include the odd proclivity of stellar nucleosynthesis—the process by which simple elements such as hydrogen and helium are transmuted into heavier elements in the hearts of giant supernovae—to yield copious quantities of carbon, the chemical epicenter of life as we know it.

As British astronomer Fred Hoyle pointed out, in order for carbon to exist in the abundant quantities that we observe throughout the cosmos, the mechanism of stellar nucleosynthesis must be exquisitely fine-tuned in a very special way.

Yet another bio-friendly feature of the cosmos is the physical dimensionality of our universe: Why are there just three extended dimensions of space, rather than one or two, or even the 10 spatial dimensions contemplated by M-theory? As has been known for more than a century, in any other dimensional set-up, stable planetary orbits would be impossible and life would not have time to get started before planets skittered off into deep space or plunged into their suns.

For centuries, it seemed that the dimensionality of the universe—three dimensions of space plus one dimension of time—was a matter of axiomatic truth, rather similar to the propositions of geometry. That was before the birth of superstring theory, and its successor, M-theory. I am going to get into M-theory more deeply in a moment, but for now I want to highlight its insistence on the fact that there are, in fact, 10 dimensions of space, and one dimension of time. The mystery is why only three of the spatial dimensions got inflated into cosmic proportions by the Big Bang, while the remaining seven stayed inconceivably minuscule. If anything else had happened—if only two spatial dimensions had been inflated, or if four had been inflated—then the universe would not have been set up to allow the emergence of life and mind as we know them.

Collectively, this stunning set of coincidences render the universe eerily fit for life and intelligence. And the coincidences are built into the fundamental fabric of our reality. As British Astronomer Royal Sir Martin Rees says, "There are deep connections between stars and atoms, between the cosmos and the microworld.... Our emergence and survival depend on very special 'tuning' of the cosmos."[8] Or, as the eminent Princeton physicist John Wheeler put it, "It is not only that man is adapted to the universe. The universe is adapted to man. Imagine a universe in which one or another of the fundamental dimensionless constants of physics is altered by a few percent one way or the other? Man could never come into being in such a universe."[9]

Scientists have been aware of this set of puzzles for decades and have given it a name—the *anthropic cosmological principle*—but there is a new urgency to the quest for a plausible explanation because of two very recent discoveries: the first at nature's largest scale, and the second at its smallest.

The first was the discovery of dark energy, which resulted from the observations of supernovae at extreme distances. Contrary to all expectations, the evidence showed that the expansion of the universe was speeding up, not slowing down. No one knows what is causing this phenomenon, although speculative explanations such as leakage of gravity into extra unseen dimensions are beginning to show up in the scientific literature.

But for our purposes, what is particularly puzzling is why the strength of dark energy—which the new Wilkinson microwave probe has revealed to be the predominant constituent of our cosmos—is so vanishingly small, yet not quite zero. If it were even slightly stronger, the universe would have been emptied long ago, scrubbed clean of stars and galaxies well before life and intelligence could evolve.

The second discovery occurred in the realm of M-theory, whose previous incarnation was known as superstring theory. If you have read Brian Greene's book *The Elegant Universe*[10] or watched the NOVA series based on it, you will know that M-theory posits that subatomic particles such as quarks, electrons, and neutrinos are really just different modes of vibration of tiny one-dimensional strings of energy. But what is truly strange about M-theory is that it allows a vast landscape of possible vibration modes of superstrings, only a tiny fraction of which correspond to anything similar to the sub-atomic particle world we observe and that is described by what is known as the Standard Model of particle physics.

Just how big is this landscape of possible alternative models of particle physics allowed by M-theory? According to Stanford physicist and superstring pioneer Leonard Susskind, the mathematical landscape is horrifyingly gigantic, permitting 10^{500} different and distinct environments, none of which appears to be mathematically favored, let alone foreordained by the theory. And in virtually none of those other mathematically permissible environments would matter and energy have possessed the qualities that are necessary for stars, galaxies, and carbon-based living creatures to have emerged from the primordial chaos.

This is, as Susskind suggests, an intellectual cataclysm of the first magnitude, because it seems to deprive our most promising new theory of fundamental physics—M-theory—of the power to uniquely predict the emergence of anything remotely resembling our universe. As Susskind puts it, the picture of the universe that is emerging from the deep mathematical recesses of M-theory is not an elegant universe at all! It's a Rube Goldberg device, cobbled together by some unknown process in a supremely improbable manner that just happens to render the whole ensemble miraculously fit for life. In the words of University of California theoretical physicist Steve Giddings, "No longer can we follow the dream of discovering the unique equations that predict everything we see, and writing them on a single page. Predicting the constants of nature becomes a messy environmental problem. It has the complications of biology."[11] Note the key word Giddings uses—*biology*—because we will be coming back to it shortly.

This really is, as Brian Greene says, the deepest problem in all of science. It really is, as Paul Davies says, the biggest of the big questions: Why is the universe life-friendly?

Three Roads to a Bio-Friendly Cosmos

If we put to one side theological approaches to this ultimate issue, what rational pathways forward are offered from the scientific community? I suggest that three basic approaches are available. Two are familiar, whereas the third is radically novel.

The first approach is to continue searching patiently for a unique final theory—something similar to $E = mc^2$—which might yet, against the odds, emerge from M-theory or one of its competitors (such as loop quantum gravity) aspiring to the status of a Theory of Everything. This is the fond hope of virtually every professional theoretical physicist, including those who have been driven to desperation by the horrendously messy and complex landscape of theoretically possible M-theory-allowed universes that distresses Susskind and other superstring theorists. Perhaps the laws and constants of nature—an ensemble the late New York Academy of Sciences president and physicist Heinz Pagels dubbed the cosmic code—will, in the end, turn out to be uniquely specified by mathematics, and thus subject to no conceivable variation. Perhaps the ultimate equations will someday slide out of the mind of a new colossus of physics as slickly and beautifully as $E = mc^2$ emerged from Einstein's brain. Perhaps. But it appears to be an increasingly unlikely prospect.

A second approach, born of desperation on the part of Susskind and others, is to overlay the theory of eternal inflation with an explanatory approach that has been traditionally reviled by most scientists that is known as the weak anthropic principle. The weak anthropic principle merely states in tautological fashion that because human observers inhabit this particular universe, it must perforce be life-friendly, or it would not contain any observers resembling ourselves. Eternal chaotic inflation, as stated by Russian-born physicist Andrei Linde, asserts that instead of just one Big Bang there are, always have been, and always will be, zillions of Big Bangs going off in inaccessible regions all the time. These Big Bangs create zillions of new universes constantly, and the whole ensemble constitutes a multiverse.

Now here's what happens when these two ideas—eternal chaotic inflation and the weak anthropic principle—are joined together. In each Big Bang, the laws, constants, and physical dimensionality of nature come out differently. In some, dark energy is stronger. In some, dark energy is weaker. In some, gravity is stronger. In some, gravity is weaker. This happens, according to M-theory-based cosmology, because the 10-dimensional physical shapes in which superstrings vibrate—known as Calabi-Yau shapes—evolve randomly and chaotically at the moment of each new Big Bang. The laws and constants of nature are constantly reshuffled by this process, similar to a cosmic deck of cards.

Here's the crucial part: Once in a blue moon, this random process of eternal chaotic inflation will yield a winning hand, as judged from the perspective of whether a particular new universe is life-friendly. That outcome will be pure chance—one lucky roll of the dice in an unimaginably vast cosmic crap shoot with 10^{500} unfavorable outcomes for every winning turn.

Our universe was a big winner, of course, in the cosmic lottery. Our cosmos was dealt a royal flush. Here is how Nobel laureate Steve Weinberg explained this scenario in a *New York Review of Books* essay a couple of years ago: "The expanding cloud of billions of galaxies that we call the big bang may be just one fragment of a much larger universe in which big bangs go off all the time, each one with different values for the fundamental constants."[12] It is no more a mystery that our particular branch of the multiverse exhibits life-friendly characteristics, according to Weinberg, than that life evolved on the hospitable Earth "rather than some horrid place, like Mercury or Pluto."[13]

If you find this scenario unsatisfactory—the weak anthropic principle superimposed on Andrei Linde's theory of eternal chaotic inflation—I can assure you that you are not alone. To most scientists, offering the tautological explanation that, because human observers inhabit this particular universe, it must necessarily be life-friendly or else it would not contain any observers resembling ourselves is anathema. It just sounds like giving up.

In my view, there are two primary problems with the Weinberg/Susskind approach. First, universes spawned by Big Bangs other than our own are inaccessible from our own universe, at least with the experimental techniques currently available to scientists. So the approach appears to be untestable. And testability is the hallmark of genuine science, distinguishing it from fields of inquiry such as metaphysics and theology.

Is String Theory Not Even Wrong?

The traditional hallmark of scientific inquiry is that it generates falsifiable predictions (or, at a minimum, falsifiable retrodictions). This defining feature of science underscores the necessity of maintaining a tight linkage between the work of theorists and observational evidence emerging from the real world. (The discoverable and discernible characteristics of the real world are presumably the sole legitimate subject of scientific study.) Absent the quality of falsifiability—a defining criterion of science generally attributed to the philosopher Karl Popper—even the most mathematically elaborate ruminations about the nature of reality are not really science, but rather a species of metaphysics or even theology.

Superstring theory (and its successor, M-theory) have recently come under a barrage of criticism because of string theorists' acknowledged failure to generate even a single falsifiable implication that could be either verified or refuted on the basis of experimental evidence. In *Not Even Wrong*,[14] Columbia mathematician Peter Woit mounts a full-throated assault on what he views as the dangerous pretensions of the M-theory crowd, who indignantly insist that their work constitutes genuine science despite their conceded inability to generate falsifiable predictions:

> As a general rule, scientific progress comes from a complex interaction of theoretical and experimental advances. In the course of the explanation of

> superstring theory and its history in the last chapter, the alert reader may have noticed the lack of any reference to experimental results. There is a good reason for this: superstring theory has had absolutely zero connection with experiment since it makes absolutely no predictions.[15]

Echoing a classic put-down of non-falsifiable scientific hypotheses by the famously blunt physicist Wolfgang Pauli, Woit maintains that superstring and M-theory are "not even wrong" and should be consigned, without further delay, to the trash heap of intellectual history:

> The failure of the superstring theory program must be recognized and lessons learned from this failure before there can be much hope of moving forward. As long as the leadership of the particle theory community refuses to face up to what has happened and continues to train young theorists to work on a failed project, there is little likelihood of new ideas finding fertile ground in which to grow. Without a dramatic change in the way theorists choose what topics to address, they will continue to be as unproductive as they have been for two decades, waiting for some new experimental results finally to arrive.[16]

Two factors, however, militate in favor of a more moderate, wait-and-see approach. The first is the cautionary example of Einstein, who formulated his great theories of relativity in primary reliance on what can only be called an unwavering faith in nature's underlying rationality and simplicity. Almost disdainful of the possibility that his theories might be experimentally disproven, Einstein was guided by his deep belief in nature's ultimate intelligibility and rationality. Needless to say, this belief was spectacularly vindicated. So perhaps we should cut the string theorists some additional slack out of deference to the Einstein precedent.

The second factor is more speculative. Perhaps the very non-uniqueness of the string theory landscape may be hinting at the possibility of a new style of predictive theory of the universe—a kind of quasi-biological theory with close analogies to Darwinism. This new kind of theory might conceivably qualify as the type of fresh cosmological thinking to which Roger Penrose referred in the concluding passage of *The Road to Reality*:

> It is quite likely that the 21st century will reveal even more wonderful insights than those that we have been blessed with in the 20th. But for this to happen, we shall need powerful new ideas, which will take us in directions significantly different from those currently being pursued. Perhaps what we mainly need is some subtle change in perspective—something that we all have missed....[17]

The Weinberg/Susskind approach extravagantly violates the mediocrity principle. The mediocrity principle, a mainstay of scientific theorizing since Copernicus, is a statistically based rule of thumb that, absent contrary evidence, a particular sample (Earth, for instance, or our particular universe) should be assumed to be a typical example of the ensemble of which it is a part. The Weinberg/Susskind approach flagrantly flouts the mediocrity principle. Instead, their approach simply takes refuge in a brute, unfathomable mystery—the conjectured lucky roll of the dice in a crap game of eternal chaotic inflation—and declines to probe seriously into the possibility of a naturalistic cosmic evolutionary process that has the capacity to yield a life-friendly set of physical laws and constants on a nonrandom basis. It is as if Charles Darwin, contemplating the famous tangled bank (the arresting visual image with which he concludes *The Origin of Species*), had confessed not a magnificent obsession with gaining an understanding of the mysterious natural processes that had yielded "endless forms most beautiful and most wonderful,"[18] but rather a smug satisfaction that *of course* the earthly biosphere must have somehow evolved in a just-so manner mysteriously friendly to humans and other currently living species, or else Darwin and other humans would not be around to contemplate it!

Indeed, the situation that confronts cosmologists today is eerily reminiscent of that which faced biologists before Charles Darwin propounded his revolutionary theory of evolution. Darwin confronted the seemingly miraculous phenomenon of a fine-tuned natural order in which every creature and plant appeared to occupy a unique and well-designed niche. Refusing to surrender to the brute mystery posed by the appearance of nature's design, Darwin masterfully deployed the art of metaphor to elucidate a radical hypothesis—the origin of species through natural selection—that explained the apparent miracle as a natural phenomenon.

The metaphor furnished by the familiar process of artificial selection was Darwin's crucial stepping stone. Indeed, the practice of artificial selection through plant and animal breeding was the primary intellectual model that guided Darwin in his quest to solve the mystery of the origin of species and to demonstrate in principle the plausibility of his theory that variation and natural selection were the prime movers responsible for the phenomenon of speciation. Today, a few venturesome cosmologists have begun to use the same poetic tool utilized by Darwin—the art of metaphorical thinking—to develop novel intellectual models that might offer a logical explanation for what appears to be an unfathomable mystery: the apparent fine-tuning of the cosmos.

The cosmological metaphor chosen by these theorists is life itself. What if life, they ask (in the spirit of the great Belgian biologist and Nobel laureate Christian de Duve) were not a cosmic accident but the essential reality at the very heart of the elegant machinery of the universe? What if Darwin's principle of natural selection were merely a tiny fractal embodiment of a universal life-giving principle that drives the evolution of stars, galaxies, and the cosmos itself?

This, as you may have guessed, is the headline summarizing the third (and radically novel) approach to answering the biggest of the Big Questions: Why is the universe life-friendly? It is the approach outlined at length in my first book, *Biocosm*.[19]

Before I get into this third approach in more detail, I want to say something upfront about scientific speculation. The approach I am about to outline is intentionally speculative. Following the example of Darwin, I have attempted to crudely frame a radically new explanatory paradigm well before all of the required building materials and construction tools are at hand. Darwin had not the slightest clue, for instance, that DNA is the molecular device used by all life-forms on Earth to accomplish the feat of what he called "inheritance." Indeed, as cell biologist Kenneth R. Miller noted in *Finding Darwin's God*, "Charles Darwin worked in almost total ignorance of the fields we now call genetics, cell biology, molecular biology, and biochemistry."[20] Nonetheless, Darwin managed to put forward a plausible theoretical framework that succeeded magnificently, despite the fact that it was utterly dependent on hypothesized (but completely unknown) mechanisms of genetic transmission.

As Darwin's example shows, plausible and deliberate speculation plays an essential role in the advancement of science. Speculation is the means by which new scientific paradigms are initially constructed, to be either abandoned later as wrong-headed detours or vindicated as the seeds of scientific revolutions.

Another important lesson drawn from Darwin's experience is important to note at the outset. Answering the question of why the most eminent geologists and naturalists had, until shortly before publication of *The Origin of Species*, disbelieved in the mutability of species, Darwin responded that this false conclusion was "almost inevitable as long as the history of the world was thought to be of short duration."[21] It was geologist Charles Lyell's speculations on the immense age of Earth that provided the essential conceptual framework for Darwin's new theory. Lyell's vastly expanded stretch of geological time provided an ample temporal arena in which the forces of natural selection could sculpt and reshape the species of Earth and achieve nearly limitless variation.

The central point is that collateral advances in sciences seemingly far removed from cosmology can help dissipate the intellectual limitations imposed by common sense and naïve human intuition. And, in an uncanny reprise of the Lyell/Darwin intellectual synergy, it is a realization of the vastness of time and history that gives rise to the crucial insight. Only in this instance, the vastness of which I speak is the vastness of future time and future history.

What I attempted to do in my first book, *Biocosm*, was to take seriously the magnitude of evolutionary change that might be possible over the course of future history in order to explore, in a tentative way, a possible third pathway to an answer to the biggest of the big questions. I call that third pathway the Selfish Biocosm hypothesis.

Originally presented in peer-reviewed scientific papers published in *Complexity*, *Acta Astronautica*, and the *Journal of the British Interplanetary Society*, my Selfish Biocosm hypothesis suggests that in attempting to explain the linkage between life, intelligence, and the anthropic qualities of the cosmos, most mainstream scientists have, in essence, been peering through the wrong end of the telescope. The hypothesis asserts that life and intelligence are, in fact, the primary cosmological phenomena, and that everything else—the constants of nature, the dimensionality of the universe, the origin of carbon and other elements in the hearts of giant supernovas, the pathway traced by biological evolution—is secondary and derivative. In the words of Martin Rees, my approach is based on the proposition that "what we call the fundamental constants—the numbers that matter to physicists—may be *secondary consequences* of the final theory, rather than direct manifestations of its deepest and most fundamental level."[22]

I began developing the Selfish Biocosm hypothesis as an attempt to supply two essential elements missing from a novel model of cosmological evolution put forward by astrophysicist Lee Smolin. Smolin had come up with the intriguing suggestion that black holes are gateways to new "baby universes," and that a kind of Darwinian population dynamic rewards those universes most adept at producing black holes with the greatest number of progeny. Proliferating populations of baby universes emerging from the loins (metaphorically speaking) of "mother universes" thus come to dominate the total population of the multiverse—a theoretical ensemble of all mother and baby universes. Black-hole-prone universes also happen to coincidentally exhibit anthropic qualities, according to Smolin, thus accounting for the bio-friendly nature of the "average" cosmos in the ensemble, more or less as an incidental side-effect.

This was a thrilling conjecture, because for the first time it posited a cosmic evolutionary process endowed with what economists call a utility function (a value that was maximized by the hypothesized evolutionary process, which in the case of Smolin's conjecture was black hole maximization).

However, Smolin's approach was seriously flawed. As the computer genius John von Neumann demonstrated in a famous 1948 Caltech lecture titled "On the General and Logical Theory of Automata,"[23] any self-reproducing object (mouse, bacterium, human, or baby universe) must, as a matter of inexorable logic, possess four essential elements:

1. A *blueprint*, providing the plan for construction of offspring.

2. A *factory*, to carry out the construction.

3. A *controller*, to ensure that the factory follows the plan.

4. A *duplicating machine*, to transmit a copy of the blueprint to the offspring.

In the case of Smolin's hypothesis, one could logically equate the collection of physical laws and constants that prevail in our universe with a von Neumann blueprint and the universe at large with a kind of enormous von Neumann factory. But what could possibly serve as a von Neumann controller or a von Neumann duplicating machine? My goal was to rescue Smolin's basic innovation—a cosmic evolutionary model that incorporated a discernible utility function—by proposing scientifically plausible candidates for the two missing von Neumann elements.

The hypothesis I developed was based on a set of conjectures put forward by Martin Rees, John Wheeler, Freeman Dyson, John Barrow, Frank Tipler, and Ray Kurzweil. Their futuristic visions suggests collectively that the ongoing process of biological and technological evolution was sufficiently robust, powerful, and open-ended that, in the very distant future, a cosmologically extended biosphere could conceivably exert a global influence on the physical state of the entire cosmos. Think of this idea as the Gaia principle extended universe-wide.

A synthesis of these insights led me directly to the central claim of the Selfish Biocosm hypothesis: that the ongoing process of biological and technological emergence, governed by still largely unknown laws of complexity, could function as a von Neumann controller, and that a cosmologically extended biosphere could serve as a von Neumann duplicating machine in a conjectured process of cosmological replication.

I went on to speculate that the means by which the hypothesized cosmological replication process could occur was through the fabrication of baby universes by highly evolved intelligent life forms. These hypothesized baby universes would themselves be endowed with a cosmic code—an ensemble of physical laws and constants—that would be life-friendly so as to enable life and ever more competent intelligence to emerge and eventually to repeat the cosmic reproduction cycle. Under this scenario, the physical laws and constants serve a cosmic function precisely analogous to that of DNA in earthly creatures: They furnish a recipe for the birth and evolution of intelligent life, and a blueprint, which provides the plan for construction of offspring.

I should add that if the fabrication of baby universes, which is the key step in the hypothesized cosmic reproductive cycle that I just outlined, sounds to you like outrageous science fiction, you should be aware that the topic has begun to be rigorously explored by such eminent physicists as Andrei Linde of Stanford, Alan Guth of MIT (who is the father of inflation theory), Martin Rees of Cambridge, eminent astronomer Edward Harrison, and physicists Lawrence Krauss and Glenn Starkman.

Is It Possible to Create a Baby Universe in the Lab?

When I wrote *Biocosm* and the preceding *Complexity* essay on which the book was based, it never occurred to me that it might be possible, in the near future, to experimentally test the most basic—and seemingly most outlandish—implication of the hypothesis: that it is possible to artificially create a new baby universe. Yet such an experiment is now being planned by a Japanese scientific team. As outlined in a recent scientific paper published in the prestigious journal *Physics Review D* and titled "Is it possible to create a universe out of a monopole in the laboratory?"[24] the team contends that it is possible to create a new inflationary universe out of a stable particle in the laboratory by inflating a theorized but never-yet-observed elementary particle called a magnetic monopole. Astonishingly, the team is now designing an actual experiment that will attempt to do precisely that. If the Japanese experiment succeeds (which is obviously a long shot), a central and highly controversial implication of the Selfish Biocosm hypothesis will have been validated.

This central claim of the Selfish Biocosm hypothesis offers a radically new and quite parsimonious explanation for the apparent mystery of an anthropic or bio-friendly universe. If highly evolved intelligent life is the von Neumann duplicating machine that the cosmos employs to reproduce itself—if intelligent life is, in effect, the reproductive organ of the universe—then it is entirely logical and predictable that the laws and constants of nature should be rigged in favor of the emergence of life and the evolution of ever more capable intelligence. Indeed, the existence of such propensity is a falsifiable prediction of the hypothesis.

What has been the reaction of the scientific community to my Selfish Biocosm hypothesis? As you might suspect, some mainstream scientists have commented that the ideas advanced in my book *Biocosm* are impermissibly speculative or impossible to verify. A few have hurled what scientists view as the ultimate epithet: that my theory constitutes metaphysics instead of genuine science.

On the other hand, some of the brightest and most far-sighted scientists have been extremely encouraging. John Barrow and Freeman Dyson have offered favorable comments and reviews. In particular, *Biocosm* has received outspoken endorsements from Sir Martin Rees (the UK astronomer royal and winner of the top scientific prize in the world for cosmology) and Paul Davies (the prominent astrophysicist and author and winner of the Templeton Prize).

As I continue to explore this hypothesis in the future, what will be of utmost interest to me and my sympathizers is whether it can generate what scientists call falsifiable implications. Falsifiabiliy or testability of claims, remember, is the hallmark of genuine science, distinguishing it from metaphysics and faith-based belief systems.

I believe that the Selfish Biocosm hypothesis does qualify as a genuine scientific conjecture on this ground. A key implication of the hypothesis is that the process of progression of the cosmos through critical thresholds in its life cycle, though perhaps not strictly inevitable, is relatively robust. One such critical threshold is the emergence of human-level and higher intelligence, which is essential to the scaling up of biological and technological processes to the stage at which those processes could conceivably exert an influence on the global state of the cosmos.

The conventional wisdom among evolutionary theorists, typified by the thinking of the late Stephen Jay Gould, is that the abstract probability of the emergence of anything similar to human intelligence through the natural process of biological evolution was vanishingly small. According to this viewpoint, the emergence of human-level intelligence was a staggeringly improbable contingent event. A few distinguished contrarians, such as Simon Conway Morris,

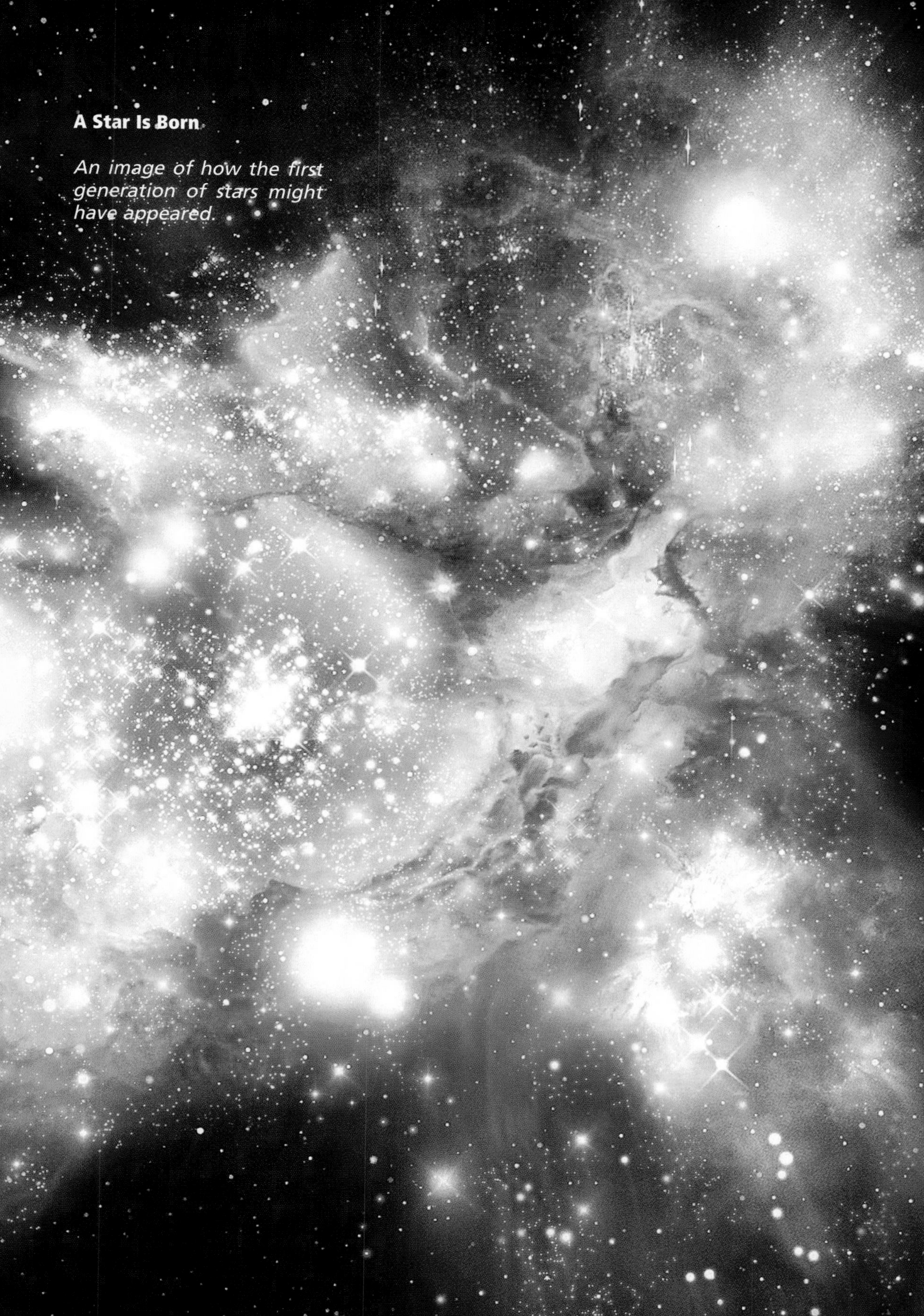

A Star Is Born

An image of how the first generation of stars might have appeared.

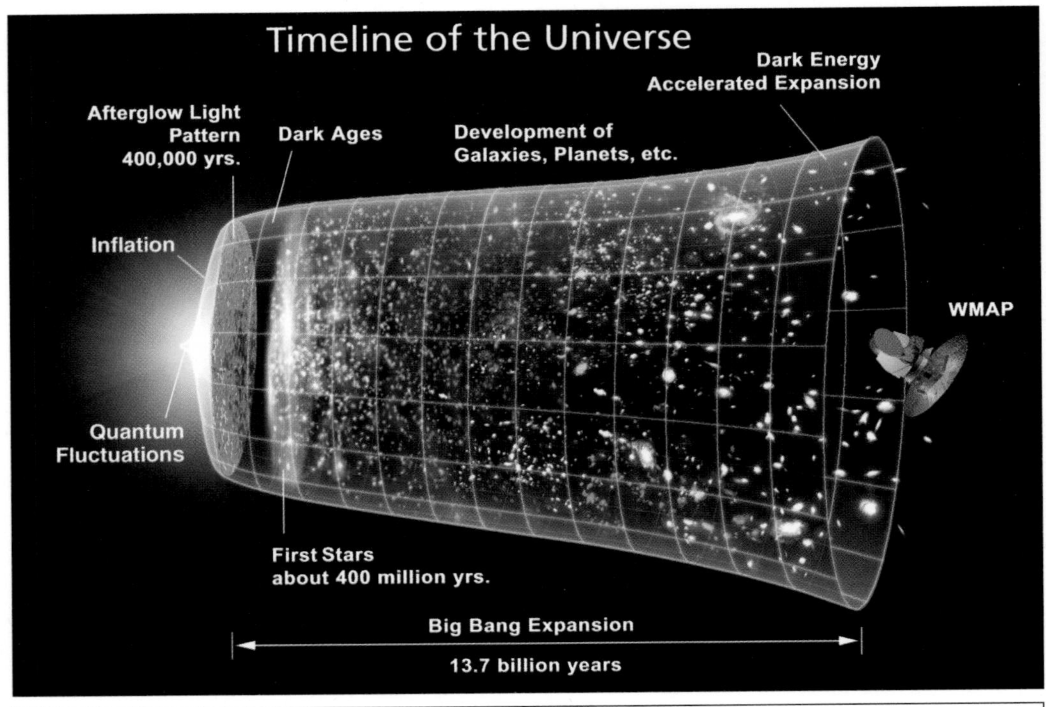
Timeline of the Universe
Dark Energy
Accelerated Expansion
Afterglow Light
Pattern
400,000 yrs.
Dark Ages
Development of
Galaxies, Planets, etc.
Inflation
WMAP
Quantum
Fluctuations
First Stars
about 400 million yrs.
Big Bang Expansion
13.7 billion years

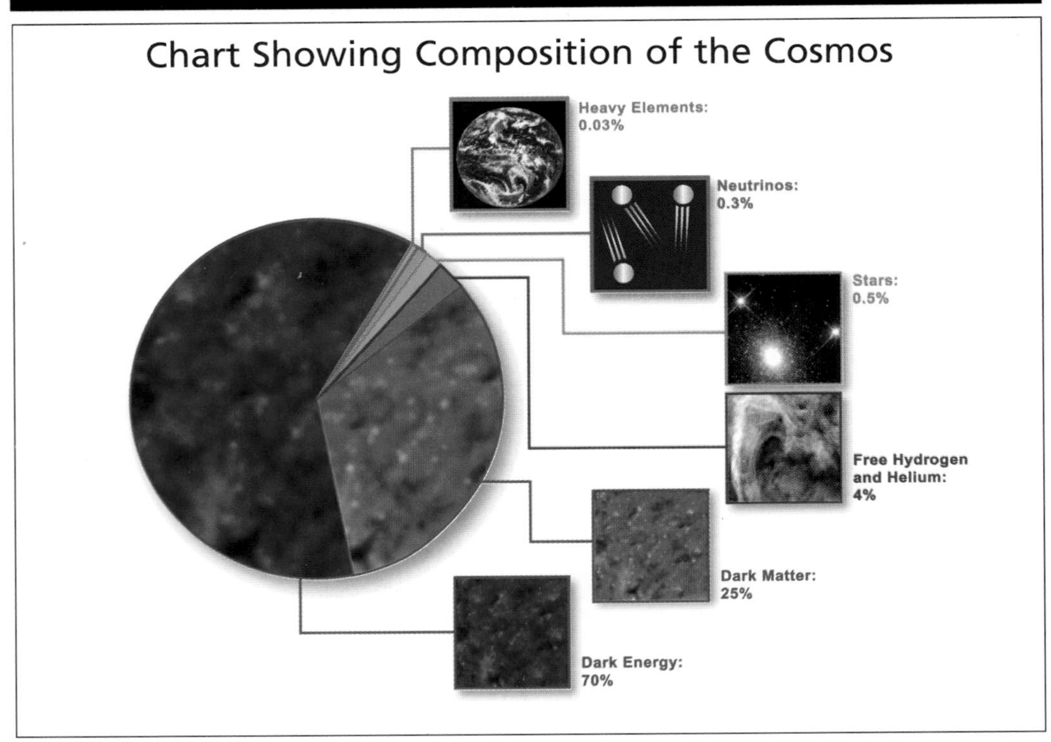
Chart Showing Composition of the Cosmos
Heavy Elements:
0.03%
Neutrinos:
0.3%
Stars:
0.5%
Free Hydrogen
and Helium:
4%
Dark Matter:
25%
Dark Energy:
70%

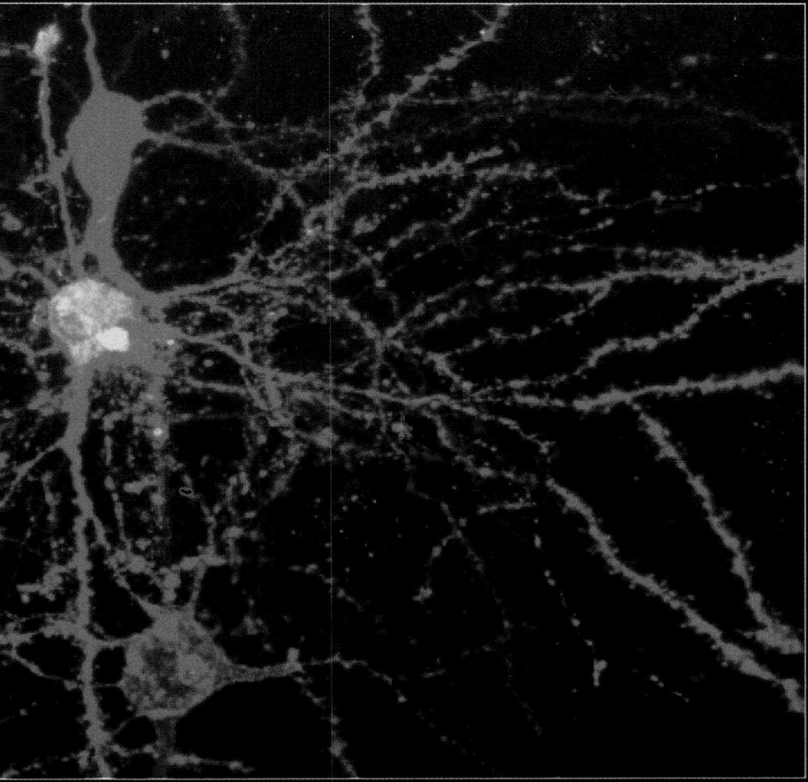

Image Courtesy of Mark N. Miller, Nelson Lab, Brandeis University

l at Birth

s the
neuron
n in the brain
le mouse have
the pattern
ribution of
the largest
e? These
the two
nouse brain
cosmos right)
distant in time,
scale show a
nd perhaps
y superficial)
ce. These
ggest that
eeply fractal,
osmological
and emergence
t of a ceaseless
a few invariant

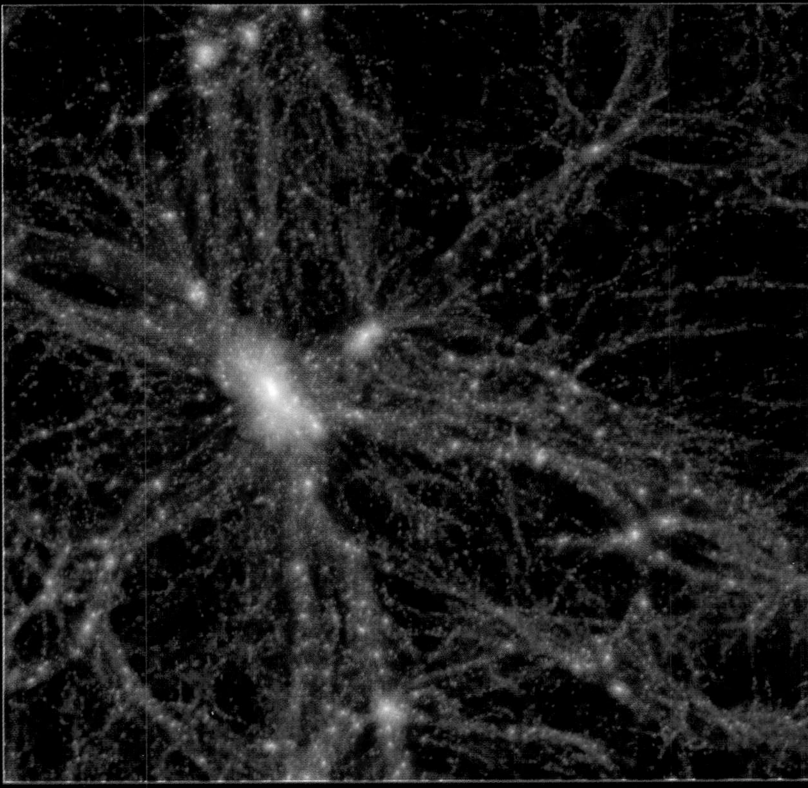

Image Courtesy of Max-Planck-Institute for Astrophysics, Garching, Germany

NASA's Conception of an Ancient Martian Ocean

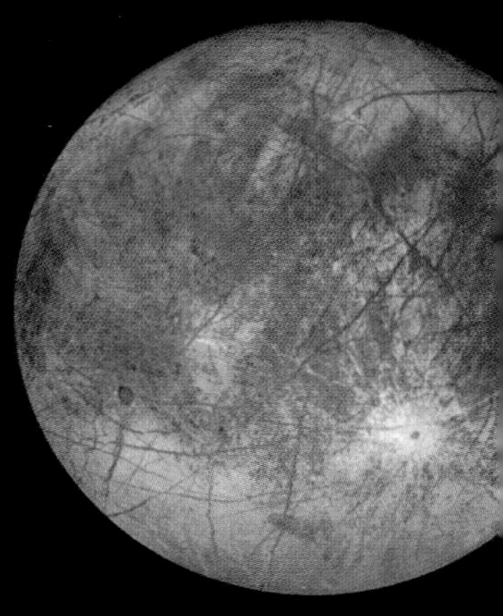

NASA Photograph of Eu

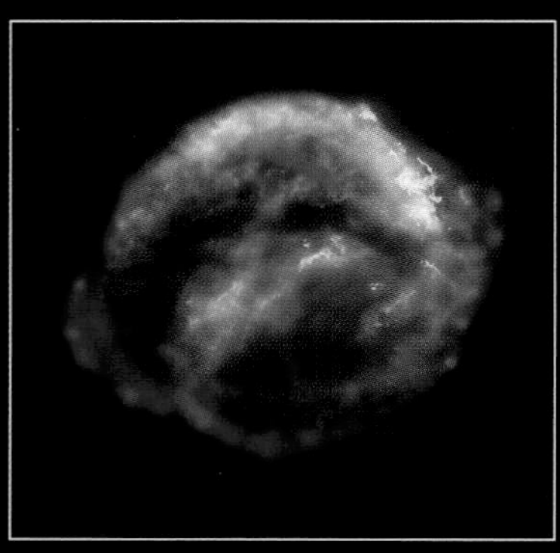

Kepler's Supernova Remnant

Supernovae are engines of chemical creation—cataclysmic cosmic forges in which all the higher elements in chemistry's periodic table are hammered into existence under conditions of enormous pressure and temperature. The image above is the remnant of a supernova explosion.

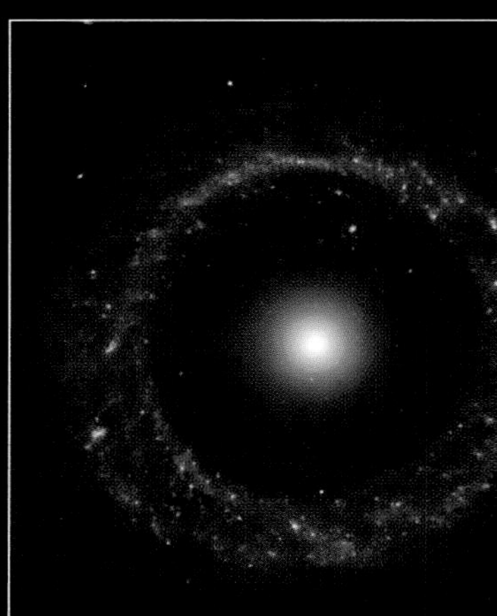

Ring of Hot Blue Stars Around Hoag's Object

If we ever discover an artifa astro-engineering, perhaps resemble this eerily symmet of stars around a mysteriou object.

The Whirlpool Galaxy

The Whirlpool Galaxy is one of the most beautiful objects in the universe, hosting numerous clusters of young and energetic stars.

The Sombrero Galaxy

Cosmic Microwave Background Radiation

Cosmic microwave backgound radiation, revealed by the WMAP probe, is the telltale fingerprint of the Big Bang.

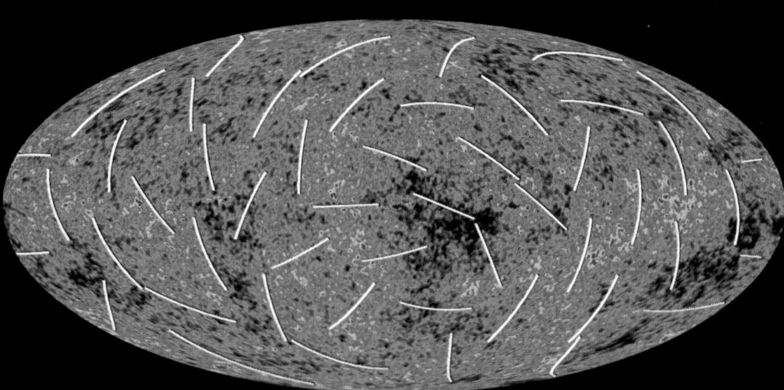

Full Sky Temperature Map of the Universe

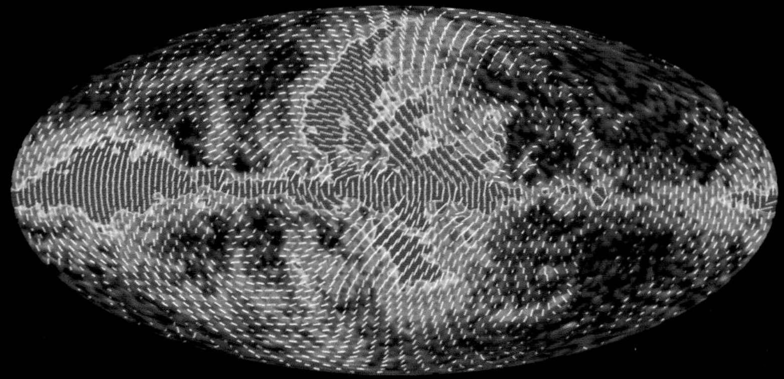

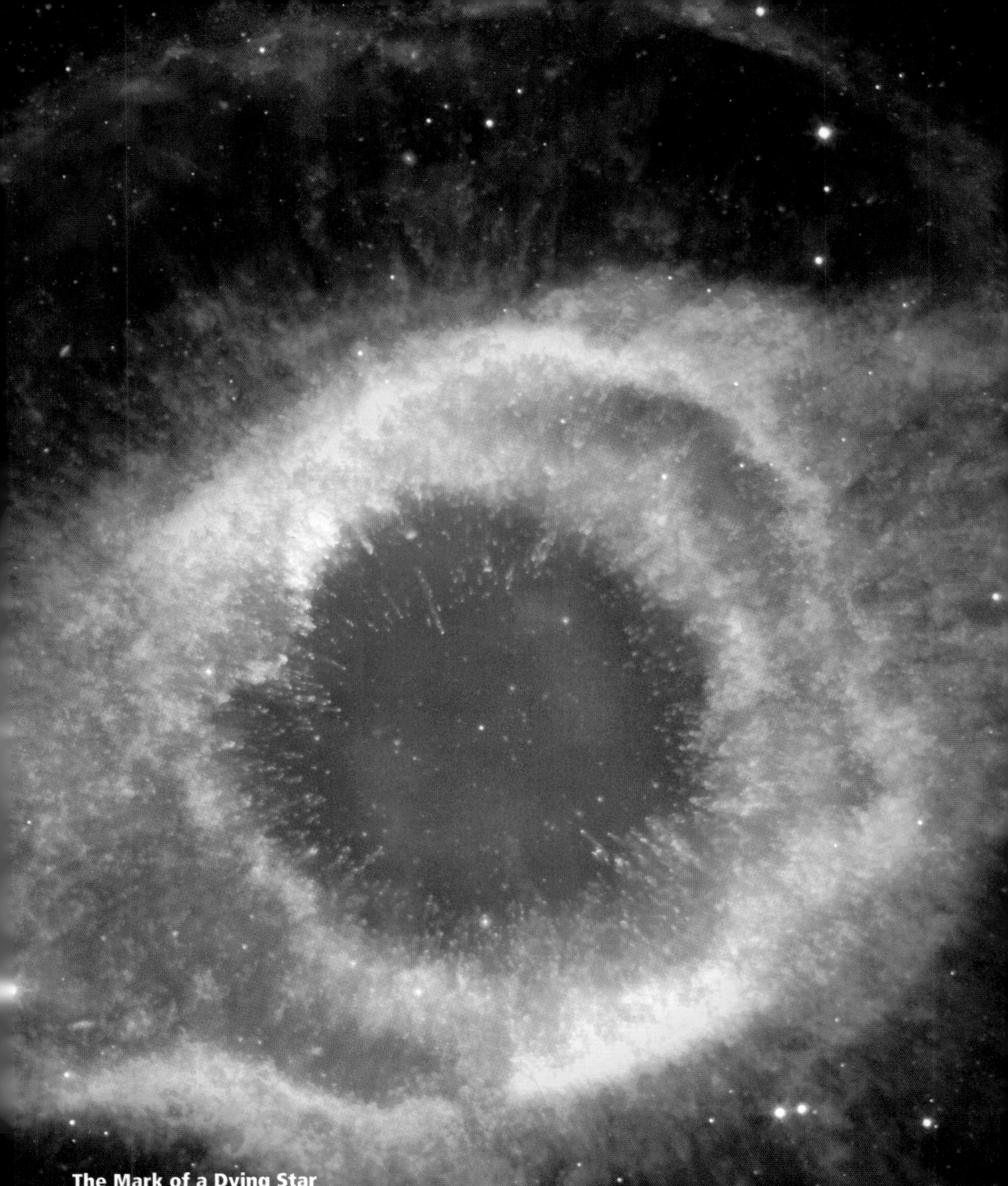

The Mark of a Dying Star

650 light-years away in the constellation Aquarius, a dying white dwarf star (about the size of Earth) is refusing to fade away peacefully. In death, it is spewing out massive amounts of hot gas and intense ultraviolet radiation, creating a spectacular object called a planetary nebula

The Cone Nebula

Infant stars are born in nebulas such as this, where they feed off the debris of prior supernova explosions.

Robert Wright, E. O. Wilson, and Christian de Duve, take an opposing position, arguing on the basis of the pervasive phenomenon of convergent evolution and other evidence that the appearance of human-level intelligence was highly probable, if not virtually inevitable. The latter position is consistent with the Selfish Biocosm hypothesis, whereas the Gould position is not.

In *Biocosm*, I suggest that the issue of the robustness of the emergence of human-level and higher intelligence is potentially subject to experimental resolution by means of at least three realistic tests: SETI research, artificial life evolution, and the emergence of transhuman computer intelligence predicted by computer science theorist Ray Kurzweil and others. The discovery of extraterrestrial intelligence, the discovery of an ability on the part of artificial life forms that exist and evolve in software environments to acquire autonomy and intelligence, and the emergence of a capacity on the part of advanced self-programming computers to attain and then exceed human levels of intelligence are all falsifiable implications of the Selfish Biocosm hypothesis because they are consistent with the notion that the emergence of ever more competent intelligence is a robust natural phenomenon. These tests don't, of course, conclusively answer the question of whether the hypothesis correctly describes ultimate reality. But such a level of certainty is not demanded of any scientific hypothesis in order to qualify it as genuine science.

A question that I am often asked is whether the Selfish Biocosm hypothesis promotes or demotes the cosmic role of humanity. Have I introduced a new anthropocentrism into the science of cosmology? If so, then you should be suspect on this basis alone of my new approach because, as Sigmund Freud pointed out long ago, new scientific paradigms must meet two distinct criteria to be taken seriously: They must reformulate our vision of physical reality in a novel and plausible way, and, equally important, they must advance the Copernican project of demoting human beings from the centerpiece of the universe to the results of natural processes.

At first blush, the Selfish Biocosm hypothesis may appear to be hopelessly anthropocentric. Freeman Dyson once famously proclaimed that the seemingly miraculous coincidences exhibited by the physical laws and constants of inanimate nature—factors that render the universe so strangely life-friendly—indicated to him that "the more I examine the universe and study the details of its architecture, the more evidence I find that the universe in some sense must have known that we were coming."[25] This strong anthropic perspective may seem uplifting and inspiring at first, but a careful assessment of the new vision of a bio-friendly universe revealed by the Selfish Biocosm hypothesis yields a far more sobering conclusion.

To regard the pageant of life's origin and evolution on Earth as a minor subroutine in an inconceivably vast ontogenetic process through which the universe prepares itself for replication is scarcely to place humankind at the epicenter of creation. Far from offering an anthropocentric view of the cosmos, the new perspective relegates humanity and its probable progeny species (biological or mechanical) to the functional equivalents of mitochondria—formerly free-living bacteria whose special talents were harnessed in the distant past when they were ingested and then pressed into service as organelles inside eukaryotic cells.

The essence of the Selfish Biocosm hypothesis is that the universe we inhabit is in the process of becoming pervaded with increasingly intelligent life—but not necessarily human, or even human-successor, life. Under the theory, the emergence of life and increasingly competent intelligence are not meaningless accidents in a hostile, largely lifeless cosmos but are at the very heart of the vast machinery of creation, cosmological evolution, and cosmic replication. However, the theory does not require or even suggest that the life and intelligence that emerge be human or human-successor in nature.

The hypothesis simply asserts that the peculiarly life-friendly laws and constants that prevail in our universe serve a function precisely equivalent to that of DNA in living creatures on Earth, providing a recipe for development and a blueprint for the construction of offspring.

Finally, the hypothesis implies that the capacity for the universe to generate life and to evolve ever more capable intelligence is encoded as a hidden subtext to the basic laws and constants of nature, stitched as though it were the finest embroidery into the very fabric of our universe. A corollary—and a key falsifiable implication of the Selfish Biocosm theory—is that we are likely not alone in the universe, but are probably part of a vast—yet undiscovered—transterrestrial community of lives and intelligences spread across billions of galaxies and countless parsecs. Under the theory, we share a possible common fate with that hypothesized community: to help shape the future of the universe and transform it from a collection of lifeless atoms into a vast, transcendent mind.

A second corollary—and the inspiration for the title of this chapter—is that, if the Selfish Biocosm hypothesis is correct, then we have already received a message of sorts from ET, although we do not recognize it as such. The message, formulated and transmitted by a supremely advanced predecessor civilization at the height of its intellectual powers, would be an instruction set or algorithm consisting of the cosmic code of an entire life-friendly predecessor cosmos. Under this scenario, the elements of the message are the physical laws and dimensionless constants of nature—the basic cosmic medium. Accordingly, ET's message is nothing less and nothing more than those laws and constants, just as the message conveyed by a genome composed of DNA from one generation to the next is nothing less and nothing more than the prescription for an ontogenetic program (a plan for the complex sequence of events by which a fertilized egg is transformed into a mature adult) as well as a recipe for replication of additional members of a particular species.

The String Landscape as Genetic Alphabet: The Subtle Virtues of a Non-Unique Cosmic Code

In his classic reflection on the nature of life (*Life Itself: Its Origin and Nature*) Francis Crick contrasted the genetic code employed by every organism on Earth with the periodic table of chemistry by underscoring the arbitrariness of the former and the ineluctable cosmic universality of the latter:

> The exact nature of the genetic code is as important for biology as Mendeleev's Periodic Table of the Elements is for chemistry, but there is an important

> difference. The Periodic Table would be the same everywhere in the universe. The genetic code appears...arbitrary....[26]

Because it was both inherently random and universally pervasive throughout the terrestrial biosphere, Crick concluded that the genetic code's ubiquity on planet Earth conveyed a subtle message about the origin of life:

> If this appearance of arbitrariness in the genetic code is sustained, we can only conclude...that all life on earth arose from one very primitive population which first used it to control the flow of chemical information from the nucleic acid language to the protein language.[27]

A quarter of a century after publication of *Life Itself* we continue to marvel at the uncanny degree of terrestrial biochemical unity—a degree of unity even greater than Crick acknowledged. But ironically, the presumed universality of the laws and constants of physics (which underlie the periodic table and every other law and principle of interest to chemists) has been called into question.

The culprit responsible for this disquieting development is string theory—more precisely, the concept of a string theory landscape containing numerous low-energy vacuua that manifest a dizzying array of physical constants and dimensional set-ups, none of which appear to be mathematically favored by the underlying theory.

The physics community has reacted with predictable horror to this messy environmental problem. Physicists, after all, had hoped that string theory (and M-theory) would, in the end, yield a "brittle" unique solution, dictated by invariant mathematical principles, that would also correspond tightly to the parameters of the Standard Model. Thus would the dream of a final theory have been finally realized.

That hope has been dashed. Now string theorists, more out of desperation than conviction, have rushed to embrace the anthropic principle as the *deus ex machina* that selects our cosmic code from the googleplex of alternatives that lurk in the mathematical recesses of the theory.

But are despair and desperation the only appropriate responses to this development? Or might the very arbitrariness of the cosmic code prevalent in our universe be trying to tell us something important, much as the arbitrariness of the genetic code was saying something important to Crick?

Indeed, might that arbitrariness—the bitter pill of non-uniqueness that string theorists have been forced to swallow—possess subtle virtues of its own, perhaps including the capacity of the Standard Model to function as a genuine code? And might that code conceivably prescribe, like a kind of deep DNA, a developmental program of cosmic ontogeny?

As argued in my scientific essay published in the *International Journal of Astrobiology* in 2005 (and included in the appendix), the life-friendly laws and constants of nature are, under the Selfish Biocosm hypothesis, a special kind of bio-signature—a cryptic message from a predecessor cosmos—just as the DNA in every cell of our bodies is a meticulously encoded message from our ancestors, both near and distant.

Dawkins on Darwin, DNA, and Cosmology

It is the most famous concluding remark in the entire canon of biological science:

> There is grandeur in this view of life, with its several powers, having been originally breathed by the Creator into a few forms or into one; and that, whilst this planet has gone cycling on according to the fixed law of gravity, from so simple a beginning endless forms most beautiful and most wonderful have been, and are being, evolved.[28]

Thus did Charles Darwin conclude *The Origin of Species*, the sacrosanct text to which British evolutionary theorist Richard Dawkins paid extended homage in *The Ancestor's Tale*,[29] a picaresque tale of life's journey across the vast span of what geologists call *deep time*—the multi-billion-year expanse separating the birth of life from the emergence of modern *Homo sapiens* some 160,000 years ago.

Dawkins modeled his narrative on Chaucer's *Canterbury Tales*—the chronicle of a journey by religious pilgrims to ancient Canterbury. Chaucer's faithful pilgrims are replaced by the almost infinite variety of living species found on planet Earth. Reversing the forward chronology of actual history,

Dawkins imaginatively portrays these species as temporal voyagers trudging, flying, swimming, crawling, wriggling, and oozing backward in time toward a common origin: the commencement of life itself.

The Ancestor's Tale eloquently conveys Darwin's brilliant unifying theme: life's miraculous diversity, bracketed by its profound unity. The diversity is manifest in the extraordinary number of ways in which the process of natural selection has sculpted the bodies and life cycles of plants, animals, and bacteria inhabiting the land, the seas, the skies and the mysterious living domain beneath the Earth's surface, dubbed the *deep biosphe*re. The underlying unity inheres in the universality of life's building blocks and its coding mechanism: With extremely rare exceptions, the biochemistry of every living creature is orchestrated by the same DNA code, faithfully transcribed into the same limited set of biological amino acids that are the building blocks of proteins.

Oddly, Dawkins's ultra-Darwinist ideology blinds him to the possibility that Darwin's great insight might have relevance in the apparent fine-tuning of the physical constants of inanimate nature, which render the cosmos mysteriously hospitable to the biological phenomena his book celebrates. Indeed, Dawkins peremptorily dismisses any suggestion that there is a need for explanation:

> To some, [a life-friendly universe] means that the laws and constants must have been deliberately premeditated from the start (although it baffles me why anybody regards this as an explanation for anything, given that the problem so swiftly regresses to the larger one of explaining the existence of the equally fine-tuned and improbable Premeditator).[30]

The cosmic code transmitted at the moment of the Big Bang that created our universe links us, in a family way, to everyone and everything in the cosmos. It underscores simultaneously our centrality and our smallness in the scheme of things.

In taking full measure of the seeming miracle of a bio-friendly universe, we should obviously be skeptical of wishful thinking and "just-so" stories. But we should not be so dismissive of new approaches that we fail to relish the sense of wonder at the almost miraculous ability of science to fathom mysteries that once seemed impenetrable—a sense perfectly captured by the great British innovator Michael Faraday when he summarily dismissed skepticism about his almost magical ability to summon up the genie of electricity simply by moving a magnet in a coil of wire. As Faraday said, "Nothing is too wonderful to be true if it be consistent with the laws of nature."

If the Selfish Biocosm hypothesis is correct, it means that we are not only the spawn of stardust, but also the architects of star-laden universes yet to come. It means that physics and chemistry eerily adumbrate the details of biology in a very specific way and that the emergence of life and intelligence is a predictable climax to the impressive but lifeless symphony of inanimate nature. It means that, against all odds, the impersonal laws of nature have somehow—amazingly and miraculously—engineered their own comprehension. And, strangest of all, they have done so by catalyzing the evolution of a conscious primate on one obscure planet who dares to dream of uncovering the ultimate secrets of the entire universe.

This is beginning to sound an awfully similar to religion. And indeed the question naturally arises: What are the implications of the Selfish Biocosm hypothesis for the field of spiritual inquiry, to which we have traditionally entrusted our deepest questions about the destiny of life and the meaning of the universe?

Part III

The Emerging Mind of the Cosmos

Part II of this book took us on a wild ride from the search for ET to the possibility of sending software versions of ourselves and our civilizations zipping across the universe at light speed. The last chapter of Part II culminated in an exploration of the almost unthinkable possibility of creating a new baby universe and rendering it bio-friendly though the transmission of a unique kind of message.

In Part III, we will enter the deep waters of philosophical reflection and ponder what all of this means. We will examine what the impact on earthly religions might be if the search for extraterrestrial civilization were to finally succeed. We will ask what the snapshots of insight we have gained from our journey to the far frontiers of cosmological science may be telling us about the ultimate nature of reality. And we will come face to face with the profound mysteries lurking in that most elusive of all phenomena—the ceaseless flow of time from the moment of the Big Bang to the final cosmic curtain call when the universe ends in . . .

But hold on, let's not spoil the final surprise!

Chapter 7

Will Religion Survive Contact?

That inimitable *ur*-skeptic Michel Shermer, publisher of *Skeptic* magazine and author of the "Skeptic" column for *Scientific American*, put it best:

> I would like to immodestly propose Shermer's Last Law...: "Any sufficiently advanced extraterrestrial intelligence (ETI) is indistinguishable from God." God is typically described by Western religions as omniscient and omnipotent. Because we are far from possessing these traits, how can we possibly distinguish a God who has them absolutely from an ETI who merely has them copiously relative to us? We can't. But if God were only relatively more knowing and powerful than we are, then by definition the deity *would* be an ETI![1]

Using Shermer's Last Law as a point of departure, what can we plausibly speculate about the possible impact on religious sensibilities of the notion that the ongoing process of evolution—both terrestrial and extraterrestrial—may be poised to yield—or, in the case of an advanced but currently undetected ET civilization, may have already yielded—the emergence of forms of intelligence vastly superior to that possessed by even the smartest human beings? Or even that we may already be in possession of subtle evidence of the existence of such superior transterrestrial intelligence without ever recognizing its significance?

Believe it or not, the topic of the probable impact on religion of the discovery of extraterrestrial intelligence has generated a voluminous literature. Although that literature has yielded no consensus with regard to the religious impact of discovering

ET, it is fair to say that many serious thinkers have concluded that disruptive effects are at least possible, if not probable. As astrophysicist Paul Davies wrote in a landmark essay published in *The Atlantic Monthly* in 2003, "Truly difficult [theological] issues surround the prospect of advanced alien beings in possession of science and technology."[2] Paramount among those difficult issues is the uncomfortable prospect that any extraterrestrial civilization we may chance to encounter is likely to be far ahead of us scientifically and intellectually:

> Because our solar system is relatively young compared with the universe overall, any alien civilization the SETI researchers might discover is likely to be much older, and presumably wiser, than ours. Indeed, it might have achieved our level of science and technology millions or even billions of years ago. Just contemplating the possibility of such advanced extraterrestrials appears to raise additional uncomfortable questions for religion.[3]

Indeed, the discovery of such advanced alien intelligence would supercharge the Copernican principle—the idea that there is nothing cosmically privileged about the location of the Earth—with a whole new level of angst-inducing doubt regarding the importance of human beings in the cosmic scheme of things, thus undermining the comforting humanity-centric foundations of the Abrahamic faiths, especially Christianity. As Davies notes:

> The world's main faiths were all founded in the pre-scientific era, when Earth was widely believed to be at the center of the universe and humankind at the pinnacle of creation. As scientific discoveries have piled up over the past 500 years, our status has been incrementally diminished. First Earth was shown to be just one planet of several orbiting the Sun. Then the solar system itself was relegated to the outer suburbs of the galaxy, and the Sun classified as an insignificant dwarf star among billions. The theory of evolution proposed that human beings occupied just a small branch on a complex evolutionary tree. This pattern continued into the twentieth century, when the supremacy of our much vaunted intelligence came under threat. Computers began to outsmart us. Now genetic engineering has raised the specter of designer babies with superintellects that leave ours far behind. And we must consider the uncomfortable possibility that in astrobiological terms, God's children may be galactic also-rans.[4]

Those theologians who reassure themselves and their flocks that nothing much would change if ET's message were discovered and decoded are delusional, Davies contends. As he wrote in a paper published by the Foundation for the Future:

> Most surveys show that theologians and ministers of religion take a relaxed view of the possibility of extraterrestrials. They do not regard the prospect of contact as threatening to their belief systems. However, they are being dishonest. All the major world religions are strongly geocentric, indeed homocentric. Christianity is particularly vulnerable because of the unique position of Jesus Christ as God incarnate. Christians believe that Christ died specifically to save mankind. He did not die to save "little green men." The alternative—that God became incarnate on planet after planet—not only has an air of absurd theatricality to it, it is also a heresy in Catholicism.[5]

Jill Tarter, SETI researcher and the model for Jodie Foster's character in the movie *Contact*, agrees with Davies that the discovery of superior extraterrestrial intelligence could have a devastating effect on the world's religions, but not necessarily on the terrestrial future of religion itself. In keeping with Shermer's Last Law, Tarter has speculated that ET's message might have the effect, intended or unintended, of evangelizing earthlings and converting them to a new cosmic faith. As she wrote in a fascinating essay titled "SETI and the Religions of the Universe":

> Let us suppose that elsewhere there is a universal religion that accurately reflects the existence of God or Gods; one that also permits a long-lived, stable, technological civilization that utilizes some technology we are capable of remotely sensing. What might we expect to hear from them? If the detected technology is information-bearing, rather than an accidental proof of existence, we can expect to learn about their God(s) and themselves as well as their view of the universe and its other inhabitants. Because new information about the universe is verifiable observationally (once it has been comprehended) skeptics and true believers alike will be converted to the revealed, superior religion, even if its practices are at first repugnant.[6]

Why is Tarter so sure that ET's message would have this seemingly bizarre evangelical effect? Tarter's conclusion rests on solid historical evidence of the revolutionary power of superior scientific insights (think Copernicus and Darwin versus Ptolemy and William Paley) as well as the sad spectacle, repeated again and again in human history, of the desecration and ultimate extinction of traditional folk religions at the hands of crusading invaders and missionaries armed with new creeds and superior weaponry:

> Would our collection of diverse, terrestrial religions and the currently nonreligious really convert so readily? I think it would be very hard to prevent. The message (assumed to have been decoded, without ambiguity) will be a missionary campaign without precedent in terrestrial history. Although it probably will not contain any overt proof of the existence of God(s), it will contain much information about the universe that appears Godlike. Unlike previous revelations, this new information would be something that science and engineering will, in time, be able to digest, verify, and reproduce. This will happen everywhere, for all people, assuming (as I do) that the information will be widely distributed, because it will be impossible to systematically suppress it over time. The information will change our lives and our world view, and we will not be able to put the genie back in the bottle. We have a history of old religions being abandoned when confronted with the superior technologies of terrestrial missionaries. In the face of a demonstrably stable social organization, and superior understanding of the nature of the universe, it will be hard for humanity to resist the appeal of this universal religion and its God(s).[7]

And what if ET turns out to be an atheist or an agnostic? In Tarter's view, although this discovery would not immediately extinguish religious faith on planet Earth, it would render the future of terrestrial religion tenuous at best: "Subsequent generations of humans, who mature with the knowledge of the existence of other technologies having long histories and no apparent need for religion, will find it harder and harder to subscribe to the unique terrestrial beliefs."[8]

A final possibility—and one that has arguably manifested itself already—is that "it is possible that we shall some day detect evidence of an extraterrestrial technology, without the ability to learn anything about the technologists or their theology. They exist; we are not alone. But what happens next?"[9] What would happen next is certainly a mystery, but perhaps not an inscrutable enigma, forever opaque to human insight and the tools of science.

The Oddest of Bedfellows: Religion, ET, and Science, Oh My!

One of the strangest aspects of the human mind is its mysterious capacity to harbor simultaneously two or more contradictory worldviews. Otherwise rational people, who in their everyday lives would not dream of doubting the quotidian evidence of their senses, willingly and routinely suspend disbelief when it comes to unabashedly affirming the existence of an unseen realm, peopled by supernatural spirits to whom one can communicate pleas for merciful indulgence and intervention—pleas that are believed to be heard and at least occasionally answered.

These same individuals believe, in disconcerting numbers, that the Earth is being regularly visited by intelligent extraterrestrial beings, even though hardly any reputable scientists subscribe to this assertion. And the ET belief system coexists, at least in the minds of many of the faithful, with the tenets of traditional Christianity and—even more curiously—with a general acknowledgment of the practical utility, if not the philosophical unassailability, of the scientific enterprise.

Indeed, educational experts have been surprised by the general pattern of tolerance on the part of fundamentalist Christian communities for what would seem to be, from their strongly held religious viewpoints, the infection of young, impressionable minds with noxious strains of "scientism." As Walter Russell Mead put it in a perceptive essay in *Foreign Affairs*:

> Shocked by recent polls showing that a substantial majority of Americans reject the theory of evolution, intellectuals and journalists in the United States and abroad have braced themselves for an all-out assault on Darwinian science. But no such onslaught has been forthcoming. U. S. public opinion has long rejected Darwinism, yet even in states such as Alabama, Mississippi, and South Carolina, which have large actively Christian populations, state universities go on teaching astronomy, genetics, geology, and paleontology with no concern for religious cosmology, and the United States continues to support the world's most successful scientific community. Most evangelicals find nothing odd about this seeming contradiction. Nor do they wish to change it.... The pragmatism of U. S. culture combines with the somewhat anti-intellectual cast of evangelical religion to create a very broad public tolerance for what, to some, might seem an intolerable level of cognitive dissonance. In the seventeenth century, Puritan Harvard opposed Copernican cosmology, but today evangelical America is largely content to let discrepancies between biblical chronology and the fossil record stand unresolved.[10]

Public Belief in ETI and UFOs

Scientific historian Steven J. Dick reports in *The Biological Universe*[11] that credible polls indicate that 58 percent of highly educated Americans believe in the existence of intelligent life on other planets, and a majority believes that UFOs have visited Earth.[12]

If these polls are accurate, they cast doubt on the dire predictions by Jill Tarter and others about the probable adverse effect on terrestrial religions of the discovery of extraterrestrial intelligence. Somehow the majority of Americans—who are among the most zestfully religious folks on Earth—have managed to reconcile their good old-fashioned faith with the zany notion that alien spacecraft are zipping through our skies. Would the mere receipt of a radio signal from a distant star have a greater cultural impact than the widespread public belief in the fact of alien visitation? I doubt it.

What's going on here? Is a majority of the public certifiably schizophrenic, or is this cacophonous jumble of ideas in our heads—belief and skepticism, reason and faith, hard-nosed empiricism and giddy bursts of reason-free rapture—telling us something profound about the nature of human nature?

The first fact to consider in our inquiry is that science is very young, while religion is immensely old—at least as old as the human race. As Karen Armstrong put it in her magisterial *A History of God*,[13] an objective "study of the history of religion has revealed that human beings are spiritual animals. Indeed, there is a case for arguing that *Homo sapiens* is also *Homo religiosus*. Men and women started to worship gods as soon as they became recognizably human; they created religions at the same time as they created works of art."[14] They did so, Armstrong concludes,

> ...not simply because they wanted to propitiate powerful forces; these early faiths expressed the wonder and mystery that seem always to have been an essential component of the human experience of this beautiful yet terrifying world. Like art, religion has been an attempt to find meaning and value in life, despite the suffering that flesh is heir to.[15]

The second fact about religion is that it is a consummate shapeshifter—enormously flexible and supremely capable, like life itself, of evolving in unforeseen directions to meet new challenges. As Armstrong puts it,

> There have been many theories about the origin of religion. Yet it seems that creating gods is something that human beings have always done. When one religious idea ceases to work for them, it is simply replaced. These ideas disappear quietly, like the Sky God, with no great fanfare.[16]

This phenomenon is not confined to the distant past. New religions—Christian Science and the Mormon faith, to name just two—emerged in the recent past, and many more are doubtless waiting in the wings of history.

The third fact is bound up with the idea of sociobiology—the notion that there is a genetically prescribed and evolutionarily derived mental scaffolding that guides our behavior and the formation of our ideas and cultural constructs. As Harvard scientist and sociobiology pioneer E. O. Wilson explained in his book *Consilience*:

> Culture is created by the communal mind, and each mind in turn is the product of the genetically structured human brain. Genes and culture are therefore inseverably linked. But the linkage is flexible, to a degree still mostly unmeasured. The linkage is also tortuous: Genes prescribe epigenetic rules, which are the neural pathways and regularities in cognitive development by which the individual mind assembles itself. The mind grows from birth to death by absorbing parts of the existing culture available to it, with selections guided through epigenetic rules inherited by the individual brain.[17]

Because these epigenetic rules—characterized by Wilson as the full range of inherited regularities of development in anatomy, physiology, cognition, and behavior as well as the algorithms of growth and differentiation that create a fully functioning organism—are genetically prescribed, they are also, according to the precepts of Darwinism, the product of evolution and natural selection:

> Some individuals inherit epigenetic rules enabling them to survive and reproduce better in the surrounding environment and culture than individuals who lack those rules, or at least possess them in weaker valence. By this means, over many generations, the more successful epigenetic rules have spread through the population along with the genes that prescribe the rules. As a consequence the human species has evolved genetically by natural selection in behavior, just as it has in the anatomy and physiology of the brain.[18]

The Sociobiology Wars: Nature Versus Nurture

The basic premise of sociobiology—that human behavioral and cultural tendencies are genetically prescribed products of the evolutionary process—was intensely opposed by prominent biologists such as Stephen Jay Gould and Richard Lewontin when the new science was invented by E. O. Wilson. It remains controversial to this day.

The controversy reflects two warring conceptions about the nature of the human psyche. The first is that the mind constitutes a blank slate at birth–a *tabula rasa* on which culture and experience can inscribe any conceivable set of habits, predilections, and abilities. The more extreme proponents of this view, such as classic Marxists, believe that humanity as a whole is a mound of raw material waiting to be shaped into whatever design is deemed optimal by despotic governments. As Lenin's admirer Maxim Gorky put it with horrifying clarity, "The working classes are to Lenin what minerals are to the metallurgist."[19]

The second conception is that the mind possesses a high degree of innate structure that has been shaped over the millennia by the forces of evolution. Under extreme versions of this view, genetic differences predispose some individuals to lives of crime and others to careers of high achievement.

The most sophisticated proponent of the second point of view is Steven Pinker, an MIT cognitive scientist. In *The Blank Slate: The Modern Denial of Human Nature* Pinker offered a masterful defense of the nature side of the age-old nature-versus-nurture debate. As he put it:

> This book is based on the estimation that whatever the exact picture turns out to be, a universal complex human nature will be part of it. I think we have reason to believe that the mind is equipped with a battery of emotions, drives, and facilities for reasoning and communicating, and that they have a common logic across cultures, are difficult to erase or redesign from scratch, were shaped by natural selection acting over the course of human evolution, and owe some of their basic design (and some of their variation) to information in the genome.[20]

This view seems so straightforward that it is difficult to believe it would stir controversy. However, as Pinker reveals, the idea that there is such a thing as an innate human nature is inflammatory because it is fundamentally at odds with what is known as the Standard Social Science Model. Under this model, only social institutions and culture are responsible for molding human thought and behavior; biology and evolution purportedly play no role whatsoever.

The standard model is the product of politics, not science. Fashioned during the early 20th century to serve as the foundation for a new societal vision of equal opportunity, blank slate orthodoxy has been pressed into community service to refute obnoxious notions of innate differences between the races and the sexes. The great dilemma is how to reconcile our commitment to the important goals of equality and social justice with the growing body of virtually irrefutable evidence that the blank slate vision of human nature is simply wrong.

The answer, Pinker suggests, is to acknowledge that fundamental values, such as human rights and a commitment to equality, must be placed on a firmer foundation than the scientifically precarious notion that the human psyche is infinitely malleable and that an innate human nature does not exist:

> The fear of the terrible consequences that might arise from a discovery of innate differences has...led many intellectuals to insist that such differences do not exist–or even that human nature does not exist, because if it did, innate differences might be possible. I hope that once this line of reasoning is laid out, it will immediately set off alarm bells. We should not concede that *any* foreseeable discovery about humans could have such horrible implications. The problem is not with the possibility that people might differ from one another, which is a factual question that could turn out one way or the other. The problem is with the line of reasoning that says that if people do turn out to be different, then discrimination, oppression, or genocide would be OK after all.[21]

Applying the tools of sociobiology to the phenomenon of religion, Michael Shermer has concluded that religion is a product of gene/culture coevolution that emerged and persisted throughout the history of the human species because it conferred a differential survival advantage on communities of believers:

> Religion is a social institution that evolved as an integral mechanism of human culture to create and promote myths, to encourage altruism and reciprocal altruism, and to reveal the level of commitment to cooperate and reciprocate among members of the community. That is to say, religion evolved as the social structure that enforced the rules of human interactions before there were such institutions as the state or such concepts

> as laws and rights.... The principal social institution available to facilitate cooperation and goodwill was probably religion. An organized establishment with rules and morals, with a hierarchical structure so necessary for social primates, and with a higher power to enforce the rules and punish their transgressors, religion evolved as the penultimate effort of these pattern-seeking, storytelling, mythmaking animals.[22]

If Shermer's biocultural theory of religion is correct, it means that the propensity to believe in a deity and to participate in organized religion (or in some cultural substitute, such as the Communist Party) is deeply rooted in human nature and thus likely to be far more resistant to extirpation by the discovery of ETI than Jill Tarter imagines. Nonetheless, it seems clear that contact with extraterrestrial intelligence will require revision of the sacred narratives of the Abrahamic faiths, especially Christianity—and perhaps even spawn entirely new religions. As Templeton Prize winner Arthur Peacocke—who is both an accomplished scientist and an ordained priest in the Church of England—has remarked with brutal honesty:

> Christians have to ask themselves (and skeptics will certainly ask *them*), what can the cosmic significance possibly be of the localized, terrestrial event of the existence of the historical Jesus? Does not the mere possibility of extraterrestrial life render nonsensical all the superlative claims made by the Christian church about his significance? Would ET, Alpha-Arcturians, Martians, et al., need an incarnation and all it is supposed to accomplish, as much as *Homo sapiens* on planet Earth? Only a contemporary theology that can cope convincingly with such questions can hope to be credible today.[23]

More broadly, Peacocke believes that "[a]ny theology, any attempt to relate God to all-that-is, will be moribund and doomed if it does not incorporate [the] perspective [provided by scientific theories like evolution and Big Bang cosmogenesis] into its very bloodstream."[24]

The fourth and final fact that may be useful in developing a provisional answer to the conundrum with which we began—the curious capacity of the human mind to simultaneously harbor inconsistent points of view regarding science, religion, and ET—is what might be called the meme theory of consciousness.

The term *meme* was coined by evolutionary theorist Richard Dawkins to denote a unit of cultural transmission functionally analogous to a gene. In a chapter of *The Oxford Encyclopedia of Evolution* authored by Daniel C. Dennett (and included as an appendix in Dennett's 2006 book *Breaking the Spell*) a meme is characterized as a "new replicator" that duplicates itself in a cultural environment.[25] As defined in the *Oxford English Dictionary*, a meme is "an element of culture that may be considered to be passed on by non-genetic means."[26]

In a 1995 essay for *WIRED* magazine, I characterized a meme-centered vision of culture and consciousness this way:

> What if culture—even consciousness itself—were nothing more than an artifact of the interaction of selfish memes, ideas capable of replicating and co-evolving with supreme indifference to their impact on human hosts?
>
> A meme-centered paradigm of human culture and consciousness is, to say the least, disconcerting. In *Consciousness Explained*, cognitive theorist Daniel Dennett captures the horror graphically: "I don't know about you, but I'm not initially attracted by the idea of my brain as a sort of dung heap in which the larvae of other people's ideas renew themselves, before sending out copies of themselves in an informational Diaspora. It does seem to rob my mind of its importance as both author and critic. Who's in charge, according to this vision—we or our memes?" A meme-focused vision of culture and consciousness acknowledges forthrightly that memes are not mere random effluvia of the human experience but powerful control mechanisms that impose a largely invisible deep structure on a wide range of complex phenomena—language, scientific thinking, political behavior, productive work, religion, philosophical discourse, even history itself.[27]

Memes, in short, follow their own simple and ruthless logic, and the logic is that of replication and survival. As Dennett has pointed out, the memes of religion incorporate an ingenious survival strategy: They are designed to be systematically immune to disproof:

> [Religion includes] a special subset of cultural items [sequestered] behind the veil of *systematic* invulnerability to disproof—a pattern found just about everywhere in human societies. As many have urged...this division into the propositions that are *designed* to be immune to disconfirmation and all the rest looks like a hypothetical joint at which we could well carve nature. Right here, they suggest, is where (proto-)science and (proto-)religion part company.... The postulation of invisible, undetectable effects that (unlike atoms and germs) are *systematically* immune to confirmation or disconfirmation is so common in religions that such effects are sometimes taken as definitive.[28]

According to the meme theory of consciousness, a fit meme does not need to be logically consistent with other memes that may succeed in a Darwinian struggle for cultural survival. To succeed, a meme only needs to be adept at replicating itself within the cultural milieu. Religious memes, in particular, have evolved a neat trick for avoiding decimation on the basis of the external evidence of the senses and logical refutation: They contain a special defense mechanism that consists of systematic immunity to disproof. It is hardly surprising, then, that the memes of religion have demonstrated extraordinary survival skills and staying power. The absence of a logical nexus between these memes and the secular teachings of science appears to be no barrier to the ongoing Darwinian success of religious ideas.

Together, these four facts—the immense age, adaptability, evolutionary origin, and proven durability of the rugged memes of religion—suggest an answer to the question posed at the beginning of this chapter: Religion is likely to survive contact with ET, but is also likely to mutate in strange and unforeseen ways as a consequence. One possible macromutation is the topic of the next section of this chapter.

Cosmotheology: A Religion for the Biological Universe

One scientist has taken the bull by the horns and argued that, far from destroying religion, the discovery of extraterrestrial intelligence might imbue the realm of the sacred with new depth, breadth, and relevance. In an essay titled "Cosmotheology: Theological Implications of the New Universe," Steven J. Dick has argued that the grand vision of nature that has come into view at the dawn of the 21st century calls for a correspondingly fundamental revision of traditional theologies.[29]

If you are imagining that Steven Dick is some hapless crank, muttering incomprehensibly as he paces across the wild fringes of legitimate science, let me enlighten you about his credentials. Dick served for many years as an astronomer and the official historian of science at the United States Naval Observatory. He was recently appointed chief historian for NASA. In this capacity, he serves as the preeminent official American chronicler of the triumphs and tragedies of the Space Age.

Although Steven Dick is a member in good standing of the space science establishment, his ideas about a new cosmotheology are—there's no other way to put it—pretty far out. For starters, Dick believes that religion needs to embrace—make that *enthusiastically* embrace—the Copernican principle at a very deep level. As he puts it, "cosmotheology must take into account that humanity is in no way physically central in the universe; we are located on a small planet around a star on the outskirts of the Milky Way galaxy."[30] Scientists have acknowledged this reality for centuries, yet the world's major religions continue to cling doggedly (and dogmatically) to their outmoded anthropocentric focus—a relic of the pre-Copernican era.

The Antiquity of the Extraterrestrial Life Debate

It may come as a bit of a surprise to learn that the debate over the possible existence of extraterrestrial life predates Christianity by centuries. As historian of science Michael J. Crowe notes in *The Extraterrestrial Life Debate: 1750-1900*:

> The question of extraterrestrial life, rather than having arisen in the twentieth century, has been debated almost from the beginning of recorded history. Between the fifth-century B.C. flowering of Greek civilization and 1917, more than 140 books and thousands of essays. reviews, and other writings had been devoted to discussing whether or not other inhabited worlds exist in the universe.[31]

Even more interesting is the fact that this scholarly debate reflects a belief, widely shared by educated elites from many historical eras, that we share the universe with extraterrestrial beings. As Crowe states:

> Moreover, as documented in this book, the majority of educated persons since around 1700 have accepted the theory of extraterrestrial life and in numerous instances have formulated their philosophical and religious positions in relation to it.[32]

Indeed, traditional theologians' obsolete anthropocentrism was the basis for a stinging critique of Christianity by that arch-patriot Thomas Paine, who wrote in his *Age of Reason* in 1793:

> To believe that God created a plurality of worlds at least as numerous as what we call stars, renders the Christian system of faith at once little and ridiculous and scatters it in the mind like feathers in the air. The two beliefs cannot be held together in the same mind; and he who thinks that he believes in both has thought but little of either.[33]

As Dick summarizes Paine's sarcastic refutation of Christian orthodoxy:

> With millions of worlds under his care, Paine argued, could we really believe that the Messiah came to save human beings on this small world? Or did the redeemer hop from one world to the next, when the number of worlds was so great that he would be forced to suffer "an endless succession of death, with scarcely a momentary interval of life"?[34]

Second, the Copernican way of thinking must be extended into the realm of biology:

> Cosmotheology must take into account that humanity probably is not central biologically.... Uniqueness of [human] form [resulting from terrestrial evolution] does not make us central to the story of the universe. Nor, one would think, should it make us the special object of attention of any deity.[35]

Third, cosmotheologists must forthrightly acknowledge that the human flock is probably not at the head of the cosmic class in terms of brainpower:

> Cosmotheology must take into account that humanity is most likely somewhere near the bottom, or at best midway, in the great chain of intelligent beings in the universe. This follows from the age of the universe and the youth of our species. The universe is in excess of ten billion years old. The genus *homo* evolved only two million years ago, and archaic *Homo sapiens* only 500,000 years ago. *Homo sapiens sapiens* is considerably younger than that, and terrestrial civilization and history cover only a few millennia. Even taking into account that the universe needed billions of years to generate the ingredients for life, if nature does select for intelligence, it has probably been doing so at numerous sites long before we arrived on the scene. Surely this has relevance to the question of our relation to any universal deity.[36]

Fourth, cosmically minded preachers and aspiring cosmotheological prophets "must be open to radically new conceptions of God, not necessarily the God of the ancients, nor the God of human imagination, but a God grounded in cosmic evolution, the biological universe, and the three principles stated above."[37]

Here Dick is saying something profoundly antithetical to traditional theologians. He is asserting that we need to discard the idea that God (or gods) exist(s) in a supernatural realm. For Dick, the god(s) of cosmotheology must be confined to the naturalistic universe that every other life form inhabits and, further, his/her/their powers, however impressive, must be exercised within the everyday world of cause and effect. In other words, Dick's god(s) are just like you and me, only stronger, faster, and smarter.

And what's wrong with this idea? As Dick points out:

> Why, we may well ask, could God not be natural? Although this raises the specter of pantheism, the natural God we have in mind is not the God of Spinoza for whom God was indwelling in nature. Our natural God is compatible with the concept of Einstein, for whom God "does not play dice" nor concern himself with the fate and actions of men. But Einstein's God "appears as the physical world itself, with its infinitely marvelous structure operating at atomic level with the beauty of a craftsman's wristwatch, and at a stellar level with the majesty of a massive cyclotron."[38]

For a sense of how deeply heretical Dick's ideas are to defenders of traditional faiths, consider the comment of George Coyne, the director of the Vatican Observatory, in response to Dick's ideas that the God of cosmotheology is bound by the laws of nature, just like human beings: "There is a special place in hell reserved for those who think God is not supernatural."[39]

Fifth, "[c]osmotheology must have a moral dimension, extended to include all species in the universe—a reverence and respect for life that we find difficult enough to foster on Earth."[40] Or, to paraphrase the philosophical musings of Mr. Spock in one of my all-time favorite *Star Trek* episodes, this new theological paradigm must embrace wholeheartedly the proposition that life's true miracle lies in its nearly infinite diversity. A cosmic deity presiding over a living universe populated by endlessly varied life forms must surely be assumed to revere all creatures great and small, natural and artificial, and of every conceivable form and hue—the brown, black, white, red, and yellow humans of Earth; the little green and grey men of far-off alien worlds; and the sentient machines and infomorphs roaming the vastness of interstellar space—with equal solicitude. For such a deity there would be no favoritism of any sort—no carbon chauvinism, no high-IQ bias, and certainly nothing that smacks of anthropocentrism.

One possible advantage of Dick's heretical approach is that it may permit a future fusion of the realms of science and theology. Why? Because a naturalistic God could, at least in principle, be studied using the methodology of science. As Dick puts it:

> A major effect of the concept of a natural God is that it has the capacity to reconcile science and religion. For those with a vested interest in the supernatural God of most standard religions, this may be too great a sacrifice for reconciliation. But consider the benefits. A *natural* God is an intelligence in and of the world, a God amenable to scientific methods, or at least approachable by them. A *supernatural* God incorporates

a concept all scientists reject in connection with their science. For some, this may be precisely the point: that God cannot be, and should not be, approachable by science. But for Einstein and many other scientists (perhaps expressed in a different way for the latter) "the cosmic religious feeling is the strongest and noblest motive for scientific research."[41]

Should Science Study Religion?

Should the tools of science be used to study the phenomenon of religion? Or should the domain of the sacred remain a shrouded enclave, shielded from the prying eyes and profane proddings of anthropologists, sociologists, economists, and evolutionary biologists?

That is the striking question at the heart of an important book by Dan Dennett entitled *Breaking the Spell: Religion as a Natural Phenomenon.* Dennett, a self-described "bright"—the stylish neologism signifying a person of the atheist persuasion that he and Richard Dawkins began to promote in twin op-ed essays in 2003—comes out squarely in favor of scientific scrutiny of the origin and nature of religious faith:

> It is high time that we subject religion as a global phenomenon to the most intensive multi-disciplinary research we can muster, calling on the best minds on the planet. Why? Because religion is too important for us to remain ignorant about. It affects not just our social, political, and economic conflicts, but the very meanings we find in our lives. For many people, probably a majority of the people on Earth, nothing matters more than religion. For this very reason, it is imperative that we learn as much as we can about it. That, in a nutshell, is the argument of this book.[42]

Religion has, of course, been studied previously, both from the inside (by theological scholars as diverse in viewpoint as Augustine, Emil Durkheim, and Mircea Eliade) and from the outside by pioneering investigators (such as William James). But only recently have the sophisticated techniques of modern science—statistical analysis, investigatory methodologies developed in the fields of sociobiology and evolutionary psychology, and methods used to associate genetic patterns with particular categories of behavior—been deployed to put religion under the microscope of objective, unbiased scientific analysis. Only now, in fact, do we possess the tool kit—especially the computational techniques—that will allow scientists to develop sophisticated models of the evolution of religious culture, analogous to dynamic software models of linguistic evolution and viral mutation.

The approach advocated by Dennett—forthright demystification of a domain of human experience whose very essence is mystery, irrationality, and faith—has provoked predictable opposition, some of it from surprising quarters. In a review of *Breaking the Spell* published in *The New York Review of Books*, Princeton physicist Freeman Dyson (forthrightly conceding his own pro-religion bias), chides Dennett for wearing his atheistic prejudices on his sleeve:

> My own prejudice, looking at religion from the inside, leads me to conclude that the good vastly outweighs the evil.... Without religion, the life of the country would be greatly impoverished.... Dennett, looking at religion from the outside, comes to the opposite conclusion. He sees the extreme religious sects that are breeding grounds for gangs of young terrorists and murderers, with the mass of ordinary believers giving them moral support by failing to turn them in to the police. He sees religion as an attractive nuisance in the legal sense, meaning a structure that attracts children and young people and exposes them to dangerous ideas and criminal temptations, like an unfenced swimming pool or an unlocked gun room.[43]

But the whole point of Dennett's thoughtful book is precisely that the origins, developmental pathways, and internal dynamics of religious communities and belief systems should be subjected to intense scientific investigation, not shunned mindlessly as pathologies associated with the consumption of dangerous and outmoded cultural opiates. To argue otherwise—to either dismiss the societal value proposition of religion *ab initio* or to agree with the late Stephen Jay Gould that religion and science are separate "magisteria" that should be contemplated in utter isolation and remain forever separated by a rigid *cordon sanitaire*—is not only literally irrational but also profoundly at odds with basic lessons of history. As I pointed out in my book *Biocosm*:

> The overlapping domains of science, religion, and philosophy should be regarded as virtual rain forests of cross-pollinating ideas—precious reserves of endlessly fecund memes that are the raw ingredients of consciousness itself in all its diverse manifestations. The messy science/religion/philosophy interface should be treasured as an incredibly fruitful cornucopia of creative ideas—a constantly coevolving cultural triple helix of interacting ideas and beliefs that is, by far, the most precious of all the manifold treasures yielded by our history of cultural evolution on Earth.[44]

In his classic Lowell Lectures delivered at Harvard in 1925, British philosopher Alfred North Whitehead put forward an intriguing explanation for the curious fact that European civilization alone had yielded the cultural phenomenon we know as scientific inquiry. Whitehead's theory was that "the faith in the possibility of science, generated antecedently to the development of modern scientific theory, is an unconscious derivative from medieval theology."[45] More specifically, he contended:

> The greatest contribution of medievalism to the formation of the scientific movement [was] the inexpugnable belief that every detailed occurrence can be correlated with its antecedents in a perfectly definite manner, exemplifying general principles. Without this belief the incredible labours of scientists would be without hope. It is this instinctive conviction, vividly poised before the imagination, which is the motive power of research—that there is a secret, a secret which can be unveiled.[46]

Whence this instinctive conviction that there is discoverable pattern of order in the realm of nature? The source of the conviction, in Whitehead's view, is not the inherently obvious rationality of nature, but rather a peculiarly European habit of thought—a deeply ingrained, religiously derived, and essentially irrational faith in the existence of a rational natural order. The scientific sensibility, in short, is an unconscious derivative of medieval religious belief in the existence of a well-ordered universe that abides by invariant natural laws, which can be discovered by dint of human investigation.

If Whitehead is correct, religion is not at all alien to scientific thought, but bears an ancestral relationship to the set of intellectual disciplines that define our concept of modernity. Western religion, in short, is the father of Western science. What could be more fitting, then, than for science to focus the lens of skeptical inquiry on issues relating to its own dimly understood paternity—that is to say, on religious belief, the historical source of scientists' boundless faith in the discoverable rationality of the cosmos?

In particular, a natural god could be the source of the mysterious fine-tuning of the laws and constants of physics that traditional cosmologists find so puzzling: "Such advanced intelligence could have fine tuned the physical constants, thus explaining the conundrum of the anthropic principle."[47]

What is the role of humanity in the cosmotheological scheme of things imagined by Dick? The NASA historian does not profess to know the answer to this ultimate question, but suggests two broad possibilities:

> Surely meaning and purpose in the universe would be quite different if we are its only life rather than one of many sentient races. And therefore theologies would be quite different. Human destiny would be quite different also; if we are alone, it may be our destiny to fill the universe with life. If extraterrestrial intelligence is abundant, it will be our destiny to interact with that intelligence, whether for good or ill, for life seeks out life.[48]

There is one additional possibility—a radical idea related to (but subtly different from) Dick's notion that we inhabit a biological universe that is predisposed, at a very deep level, to spew forth life and mind whenever it gets half a chance. That radical possibility is this: The universe may not be merely life-friendly and biogenesis-prone. Instead, the universe may actually *be* an emerging life-form, a developing biosphere, a kind of cosmic superorganism in the process of emergence and ontogenesis, with a built-in capacity for eventual replication. As physical chemist and philosopher Michael Polanyi put it, "This universe is still dead, but it already has the capacity of coming to life."[49] To state the proposition differently, the universe that Nobel physicist Steven Weinberg famously described as comprehensible but pointless may be governed by telic processes in the same way that a developing embryo is directed by its internal genetic programming to grow into a mature adult. In short, the mostly dead cosmos we see all around us may constitute what I have dubbed a *biocosm aborning*. And in the emerging life and ontogeny of the biocosm, we human beings may play a vital role: We may be the equivalent of the cells or even the intracellular organelles of the biocosm—its life-giving mitochondria and chloroplasts.

Chapter 8

The Universe Is Coming to Life

After *Biocosm* was published, I had the opportunity to speak to a number of audiences, both lay and scientific. One of the most enthusiastic groups I addressed was an overflow crowd that gathered on the beautiful campus of a spiritual community in Lenox, Massachusetts, known as EnlightenNext. My lecture took place on an unseasonably warm evening in the early spring of 2006. New life was sprouting everywhere.

After I delivered my speech, I answered a number of questions, most of them similar to those that I had fielded following similar presentations. But one question really stumped me, at least initially: "Can you summarize the basic idea of *Biocosm* in one sentence?" I was silent for perhaps 30 seconds, wracking my brain for a snappy answer. How could I encapsulate the detailed cosmological and evolutionary hypothesis painstakingly articulated in *Biocosm* in one sentence? And then it came to me, as I remembered the tender young plants that were just beginning to push up through the hard New England soil. "The universe is coming to life," I finally answered. "Yes, that's it. The universe is coming to life."

The universe is coming to life. Not generating living beings haphazardly as the result of a random toss of the chemical dice. Not transforming inert matter into a growing, evolving biosphere as the consequence of a spectacularly improbable cosmic accident that happened, against all odds (and perhaps only once throughout all of space and time), on an ordinary planet orbiting an undistinguished star in the outer reaches of an ordinary galaxy.

No, the universe is coming to life, purposely and in accordance with a finely tuned cosmic code that is the precise functional equivalent of DNA in the terrestrial biosphere. The universe, under this interpretation, is a kind of vast emerging organism in the process of self-assembly and self-animation, endowed with the capacity to not only replicate itself, but also to transmit heritable traits—that same cosmic code, consisting of the laws and constants of physics, which not only prescribes an ontogenetic program but, again similar to DNA, furnishes a recipe for the self-assembly of offspring (so-called baby universes).

The Selfish Biocosm Hypothesis

The Selfish Biocosm hypothesis asserts that the anthropic qualities that our universe exhibits can be explained as incidental consequences of a cosmological replication cycle in which a cosmologically extended biosphere supplies two of the essential elements of self-replication, as identified by mathematician and computer pioneer John von Neumann. Further, the hypothesis asserts that the emergence of life and intelligence are key epigenetic thresholds in the cosmological replication cycle, strongly favored by the physical laws and constants of inanimate nature. Under the hypothesis, those laws and constants function precisely as the functional counterpart to DNA: They furnish the recipe by which the evolving cosmos acquires the capacity to generate life and ever more capable intelligence. The hypothesis reconceives the process of earthly evolution as a minuscule subroutine in an inconceivably vast process of cosmic ontogenesis. A falsifiable implication of the hypothesis is that the emergence of increasingly intelligent life is a robust phenomenon, strongly favored by the natural processes of biological evolution and emergence.

Under this hypothesis, the ancient and majestic process of terrestrial Darwinian evolution is a minor subroutine in a vastly larger process of cosmic ontogeny. Moreover, in accord with the Copernican principle of mediocrity, other instances of biological evolution—other unfoldings of evolutionary subroutines—are assumed to be as common throughout the cosmos as the emergence of specialized cells in a developing human embryo.

This is the vision of the biocosm—a universe in which life and the life-yielding propensity of the physical laws and constants of nature are the central reality. As I put it in my book, *Biocosm*:

> The essence of what I am calling the "Selfish Biocosm" hypothesis is that the universe we are privileged to inhabit is literally in the process of transforming itself from inanimate to animate matter.... Under this theory, the emergence of life and intelligence are not meaningless accidents in a hostile, largely lifeless cosmos but exist at the very heart of the vast

machinery of creation, cosmological evolution, and cosmic replication.[1]

Q&A About the Selfish Biocosm Hypothesis

What is the basic idea of the Selfish Biocosm hypothesis?

The basic idea is that life and intelligence are the primary cosmic phenomena and that everything else—the constants of nature, the dimensionality of the universe, the origin of carbon and other elements in the hearts of giant supernovas, the pathway traced by biological evolution—is secondary and derivative. According to this theory, the emergence of life and intelligence are not meaningless accidents in a hostile, largely lifeless cosmos but at the very heart of the vast machinery of creation, cosmological evolution, and cosmic replication.

How does that idea fit into current mainstream thinking about the nature of the universe?

We have entered an exciting era of precision cosmology. New tools such as the Wilkinson Microwave Anisotropy Probe (WMAP) are permitting scientists to study the universe and its origin with unprecedented clarity. Though the experimental results are unambiguous, their implications are startling and disconcerting. For instance, the WMAP measurements reveal that the cosmos is composed predominantly of a mysterious dark energy that acts as anti-gravity. The problem is that no one has the slightest clue about the nature of dark energy. Another problem is that the closer we examine the details of the natural laws and constants that govern the behavior of every object in the universe, the more it appears that those laws and constants have been mysteriously fine-tuned to be life-friendly. The Selfish Biocosm hypothesis attempts to unravel this particular mystery.

What is the hidden role of the laws and constants of nature under the Selfish Biocosm hypothesis?

Under the hypothesis, the capacity of the universe to generate life and to evolve ever more capable intelligence is encoded as a hidden subtext to the basic laws and constants of nature, stitched as though the finest embroidery into the very fabric of our universe. The oddly life-friendly laws of nature that prevail in our cosmos serve a function precisely equivalent to that of DNA in living creatures on Earth, providing a recipe for development of a living universe and a blueprint for the construction of offspring (baby universes).

How does your work on this hypothesis differ from that of scientists in more traditional settings?

My role, at least as I perceive it, is not to serve as an experimentalist or a number-cruncher, but rather as a synthesizer of a host of emerging ideas now being advanced in a variety of seemingly unrelated scientific fields, including cosmology, astrobiology, M-theory, artificial life, and evolutionary theory. I have endeavored to extract key insights from these disparate disciplines in order to connect the dots and map out a plausible scientific explanation for the inexplicably life-friendly quality of the physical laws and constants that prevail in our universe—a dazzlingly improbable feature of the cosmos that poses, as Paul Davies says, "the biggest of the Big Questions." The very fact that I am an outsider from the viewpoint of the scientific establishment means that I am not captive to the paradigms and prejudices of a particular discipline. This is a weakness inasmuch as I lack the depth of expertise in cosmology possessed by traditional astrophysicists such as Andrei Linde and Neil Turok. But it is also a key advantage, because, as a synthesizer and a scientific generalist, I am able to formulate a "crude look at the whole," in the wonderful phrase of Nobel laureate Murray Gell-Mann. My approach also happens to coincide with the methodology of complexity theory, the field of science with which I am most comfortable and in which I have published extensively.

What do mainstream scientists think of the Selfish Biocosm hypothesis?

My first book, *Biocosm*, received unexpected praise from some of the top cosmologists, physicists, and mathematicians in the world, including British Astronomer Royal Sir Martin Rees, Paul Davies, John Casti, and Seth Shostak. Others have commented that the ideas advanced in the book are impermissibly speculative or impossible to verify. A few have hurled what scientists view as the ultimate epithet: that my theory constitutes "metaphysics" instead of genuine science. I beg to differ.

What prominent thinkers have advanced ideas similar or related to the Selfish Biocosm hypothesis?

Astronomer Royal Sir Martin Rees; British astrophysicist John Barrow; physicists Freeman Dyson and John Wheeler; cosmologists Lee Smolin, Paul Davies, and Frank Tipler; evolutionary theorist and Nobel laureate Christian de Duve; evolutionary biologists Lynn Margulis, Harold Morowitz, and Simon Conway Morris; complexity theorists Stuart Kauffman and Stephen Wolfram; French religious philosopher Teilhard de Chardin; popular science author Robert Wright; computer theorist Ray Kurzweil; and, to some degree, Stephen Hawking.

How does the Selfish Biocosm hypothesis treat Darwin's theory of evolution?

Under the Selfish Biocosm hypothesis, the immense saga of biological evolution on Earth is a minor subroutine in the inconceivably lengthy process through which the universe becomes increasingly pervaded with ever more intelligent life. Thus, the hypothesis does not challenge Darwinism, but seeks to place it in a cosmic context in which life and intelligence play a central role in the process of cosmogenesis. Put differently, the hypothesis reconceives the process of earthly phylogeny as a minuscule element of a vastly larger process of cosmic ontogeny.

What are the religious implications of the hypothesis?

The hypothesis is inconsistent with traditional monotheistic notions of an unknowable supernatural Creator. Freeman Dyson has famously written that the idea of sufficiently evolved mind is indistinguishable from the idea of the mind of God. The Selfish Biocosm hypothesis takes Dyson's assertion of equivalence one step further by suggesting that there is a discernible and comprehensible evolutionary ladder by means of which mortal minds will one day ascend into the intellectual stratosphere that will be the domain of superminds—what Dyson would call the realm of God. To use Dyson's terminology, the hypothesis implies that the mind of God is the natural culmination of the evolution of the mind of humans and other intelligent creatures throughout the universe, whose collective efforts conspire—admittedly, without any deliberate intention—to effect a transformation of the cosmos from lifeless dust to vital, living matter, capable of the ultimate feat of life-mediated cosmic reproduction.

Is the Selfish Biocosm hypothesis really just a religious screed in disguise—a subtle form of Creationism or Intelligent Design proselytizing similar to Michael Behe's *Darwin's Black Box*?

Definitely not. The hypothesis is adamantly and consistently naturalistic in focus. The ideas that underlie the hypothesis were originally presented in prestigious peer-reviewed scientific journals (*Complexity*, *Acta Astronautica,* and the *Journal of the British Interplanetary Society*). Indeed, the prime objective of the book is to provide the framework for a scientifically plausible and testable formulation of the strong anthropic principle—the notion that life and intelligence have not emerged in a series of random accidents, but are essentially hard-wired into the laws of physics and into a vast cycle of cosmic creation, evolution, death, and rebirth.

Are there immediate and long-term implications of the hypothesis?

I don't have a crystal ball, but I hope that it will encourage more scientists (both professionals and amateurs) to think holistically about the challenge of deciphering what the late physicist Heinz Pagels called the cosmic code—the full suite of life-friendly laws and physical constants that prevail in our universe. I hope it will provoke students of science to recall, in the spirit of Newton, that our minuscule island of scientific knowledge is surrounded by a fathomless ocean of undiscovered truth. And I hope that it will rekindle the sense of wonder at the achievements of science, so perfectly captured by the great British innovator Michael Faraday when he summarily dismissed skepticism about his almost magical ability to summon up the genie of electricity simply by moving a magnet in a coil of wire. As Faraday noted, "Nothing is too wonderful to be true if it be consistent with the laws of nature."

In the four years since *Biocosm* was published, a number of developments have occurred that lend support to this admittedly speculative hypothesis. They include a growing recognition that, as a result of advances in string theory and M-theory, we may finally have found a tentative answer to Einstein's famous question of whether God had any choice in fashioning the laws and constants of physics. The answer appears to be yes—leading directly to the question of why our stunningly life-friendly subset of laws and constants just happened to be selected from a nearly infinite set of alternatives at the moment of the Big Bang that launched the evolution of our universe.

Where, Oh Where Is Our Anthropic Cosmos on the Vast M-Theory Landscape?

M-theory has come under increasing criticism because of its failure to generate any falsifiable implications. A related critique is that the original dream of string theorists—to come up with a unique theoretical solution that would reproduce the seemingly arbitrary parameters of the Standard Model and demonstrate the mathematical inevitability of that solution—has been jettisoned by leading practitioners in favor of the landscape approach, which embraces the very diversity of potential M-theory-permitted vacuua as a virtue, and proposes that our particular corner of the landscape is special only in that it permits the existence of observers such as ourselves. In the words of M-theory critics, the landscape approach demotes the theory from a candidate Theory of Everything into a worthless "theory of anything," or even a "theory of nothing."

Perhaps the M-theorists could begin to salvage the rapidly dwindling reputation of their field by concentrating on a single humble task: demonstrating not that the cosmos we happen to inhabit is mathematically foreordained by the theory, but simply that it exists somewhere on M-theory's vast theoretical landscape.

Alas, even this chore has proven too daunting for M-theory brainiacs. As noted by *Scientific American*, the task of locating one or more string vacuua that would yield such parameters as the observed value of so-called dark energy may be too computationally difficult. As a leading researcher put it, "The old ambition of exactly finding the precise vacuum in which we live may be intractable."[2]

If string theorists cannot even *locate* our oddly bio-friendly cosmos on the M-theory landscape—let alone demonstrate its mathematical inevitability—then it seems clear that, at the very least, the theory confronts severe challenges.

A related insight is that none of the zillions of possible universes permitted by the mathematics of M-theory appears to be mathematically favored, let alone foreordained (and certainly not the bio-friendly cosmos we happen to inhabit).

There is also a realization that some physically plausible mechanism must exist for selecting our bio-friendly cosmos from the googleplex of M-theory-allowed alternatives, and a hint that a cosmic reproductive principle of some sort (perhaps fueled by Andrei Linde's hypothesized eternal inflation) is at the heart of the mysterious process that generates those alternative universes.

Finally, there has been an admission by a growing number of leading cosmologists that, assuming there is a process of eternal inflation, no plausible physical mechanism has been hypothesized that is capable of transmitting from a mother universe to its progeny the set of physical laws and constants that prevailed in the predecessor cosmos. Absent a heredity mechanism, those laws and constants are assumed by traditional cosmologists and M-theorists to evolve chaotically at the moment of each new Big Bang—to be reshuffled randomly and endlessly, as if a giant deck of cards, each time the process of eternal chaotic inflation gives birth to a new Big Bang and a new baby universe.

As a consequence of these developments and admissions, the perceived need for creative thinking about the nature and origin of our peculiarly life-friendly cosmos has greatly intensified since the publication of *Biocosm*.

At the same time, there is growing recognition of just how spectacularly and improbably bio-friendly our cosmos actually is. This startling fact is prompting leading physicists such as Stephen Hawking to push the envelope of physical theory in an almost desperate attempt to come up with a plausible explanation. Hawking's approach (which is critiqued at length in my *International Journal of Astrobiology* essay included as Appendix A) centers on the so-called many-worlds interpretation of quantum theory as applied to the universe as a whole. His notion is that the physical evolution of the cosmos constitutes a simultaneous exploration of every possible branch of the primordial cosmic quantum wave function, and that we are capable of experiencing only those branches of the wave function that are hospitable to our kind of life. As I have argued, Hawking's hypothesis appears to be untestable in principle, although Hawking and a colleague have recently contended that the cosmic microwave background radiation may contain telltale imprints of the early evolution of the cosmic wave function.

My approach to the puzzle of an improbably life-friendly cosmos differs profoundly from that of Hawking and the M-theory crowd, and rests on extrapolation from what NASA historian Steven Dick has called the biological universe worldview:

> The biological universe [is] more than an idea, more than another theory or hypothesis; it [is] sufficiently comprehensive to qualify as a worldview.[3]

The essence of the biological universe worldview—a revolutionary perspective that Dick believes will require as profound a revision in our way of thinking about the cosmos as the Copernican and Darwinian revolutions—is that "planetary systems are common, that life originates wherever conditions are favorable, and that evolution culminates with intelligence."[4] This worldview is the polar opposite of the viewpoint expressed, with eloquent existential angst, by Nobelist Jacques Monod in *Chance and Necessity*: "The universe was not pregnant with life, nor the biosphere with man."[5]

The biological universe worldview received what may have been its most eloquent endorsement from another Nobel laureate, Christian de Duve, who disagreed sharply with Monod:

> The universe is not the inert cosmos of the physicists, with a little life added for good measure. The universe *is* life, with the necessary infrastructure around; it consists foremost of trillions of biospheres generated and sustained by the rest of the universe.[6]

A key assertion of the biological universe worldview is that the properties of the inanimate physical universe are inherently bio-friendly and conducive to biological evolution. Indeed, those physical properties and processes are viewed as inseparably linked to the biological evolutionary process. As Lawrence J. Henderson, an early proponent of this view (whose contributions were largely overlooked during his lifetime), put it:

> The properties of matter and the course of cosmic evolution are now seen to be intimately related to the structure of the living being and to its activities; they become, therefore, far more important in biology than has been previously suspected.

> For the whole evolutionary process, both cosmic and organic, is one, and the biologist may now rightly regard the universe in its very essence as biocentric.[7]

Another bedrock assumption of the biological universe worldview is that the linked processes of cosmological and biological evolution are inherently open-ended and capable of feats of what has been called "macro-engineering" that we can scarcely imagine.

Cosmic Macro-Engineering as Evidence of ET

In a recent paper, Milan M. Cirkovic, a scientist with the Astronomical Observatory of Belgrade, called for a new approach to SETI and astrobiology that would concentrate on a search for artifacts of cosmic macro-engineering rather than for ET's stray or directed radio or optical signals.[8] Such artifacts might include so-called Dyson shells (named after Princeton physicist Freeman Dyson, who first conceived of them), which would entrap all the energy emitted by a star and harness it for useful purposes, as well as such exotica as artificial rings constructed around planets or stars.

One reason Cirkovic believes this is a potentially more fruitful approach than continuing to patiently search the skies for ET's radio twitters or laser flashes is that extraterrestrial civilizations, if they exist, are likely to be immensely older and more advanced than the societies of Earth: "It seems preposterous even to contemplate any possibility of communication between us and [billion-year]-older supercivilizations."[9]

The search mode advocated by Cirkovic (which he calls the Dysonian approach, in deference to Freeman Dyson), would not be hindered by the incommensurability between terrestrial technology and more advanced alien science:

> The Dysonian approach to search for other intelligent societies can be briefly summarized as follows. Even if they are not actively communicating with us, that does not imply that we cannot detect them and their astro-engineering activities. Their detection signatures may be much older than their communication signatures. Unless [advanced technological civilizations] have taken great lengths to hide or disguise their [infrared] detection signatures, the terrestrial observers should still be

able to observe them at those wavelengths and those should be distinguishable from normal stellar spectra.[10]

Coupled with this assumption is the related proposition that, as physicist Eric Chaisson has written, "the destiny of the Universe may well be determined by the life that arises from it."[11] More specifically, in the words of Princeton physicist Freeman Dyson, "Life may succeed against all odds in molding the Universe to its own purposes. And the design of the inanimate Universe may not be as detached from the potentialities of life and intelligence as scientists of the 20th century have tended to suppose."[12]

Related to this idea about the ultimate possibilities of cosmic engineering at the largest conceivable scale is the assumption that the forms of life and intelligence that will be capable of this monumental task will almost certainly be post-biological in nature. As Steven Dick has observed in the context of a critique of the biologically focused SETI search program:

> The universe over the billions of years that intelligence has had to develop will not be a biological universe, but a postbiological universe. Biologically based technological civilization...is a fleeting phenomenon limited to a few thousand years, and exists in the universe in the proportion of one thousand to one billion, so that only one in a million civilizations are biological. Such are the results of taking cultural evolution seriously, and applying the Intelligence Principle and the insights of Moravec, Kurzweil and Tipler to the entire universe.[13]

The Intelligence Principle

Steven Dick has developed what he calls the Intelligence Principle as a shorthand summary of what he regards as an intrinsic trend in cultural evolution—a trend that may be universal among biospheres scattered throughout the universe:

> I adopt what I term the central principle of cultural evolution, which I refer to as the Intelligence Principle: *the maintenance, improvement and*

perpetuation of knowledge and intelligence is the central driving force of cultural evolution, and... to the extent intelligence can be improved, it will be improved.... The Intelligence Principle implies that, given the opportunity to increase intelligence (and thereby knowledge), whether through biotechnology, genetic engineering or AI, any society would do so, or fail to do so at its own peril.[14]

In Dick's view—and in the view of the prominent futurists whom he cites, including Ray Kurzweil, Frank Tipler, Hans Moravec, and Freeman Dyson—the Intelligence Principle will lead inevitably to the emergence of post-biological intelligence.

My Selfish Biocosm hypothesis seeks to synthesize the central concepts of Dick, Dyson, Henderson, Tipler, Kurzweil, Chaisson, de Duve, and other scientists who have thought deeply about the mystery of a life-friendly cosmos. The hypothesis then goes on to propose what I contend is a physically plausible mechanism that might explain this mystery. As stated in my *International Journal of Astrobiology* paper,[15] which is attached as Appendix A, the heart of my contention is that at the culmination of the cosmic evolutionary process, billions and billions of years from now, highly evolved life and mind will play a central role in reproducing our cosmos. So, too, I hypothesize, the peculiarly life-friendly laws and physical constants that prevail in our universe will play a key role in the process of cosmic replication: They will serve a function equivalent to that of DNA in the terrestrial biosphere, furnishing a recipe for the birth and growth of a new life-friendly baby universe as well as a blueprint that will allow the Big Baby to eventually reproduce itself and give birth to its own life-friendly cosmic progeny in the fullness of future time. The Selfish Biocosm hypothesis actually predicts (technically, *retrodicts*, in the jargon of science) that the physical laws and dimensionless constants prevailing in our cosmos will be life-friendly. Equally important, the hypothesis, if correct, means that those life-friendly laws and constants will predictably generate life and advanced intelligence just as the DNA of elephants, whales, humans, and every other species on planet Earth will reliably generate individual organisms that are members of those species. This implies that the bio-friendliness of our cosmos is not, in fact, an astronomically improbable random accident, but rather a cosmic-scale biosignature, analogous to the presence of free oxygen in Earth's atmosphere, which is a terrestrial-scale biosignature.

The Selfish Biocosm hypothesis rests in part on a kind of unique *existence proof* of the power of life to transform insert matter into living, breathing ecosystems—Exhibit A in support of the hypothesis is the very existence of the earthly biosphere—and in part on a presumed absence of limits on the capacity of living matter and mind to infiltrate and effect a transformation of the cosmos. In the spirit of Darwin, who considered the process of artificial selection to be an appropriate metaphor with which to illuminate the power of natural selection to transform a few primordial organisms into the variegated biosphere we observe all around us, I consider the demonstrated transformational power of life on earth as a suitable metaphor with which to illuminate the ultimate capacity of life and mind to transform the mostly dead universe into vital, living matter.

Existence Proofs

Tendering an existence proof is a foolproof way to win any scientific argument. For instance, if someone were to produce an actual specimen of the Loch Ness monster or Sasquatch, that irrefutable evidence would constitute an *existence proof* of the conjecture that such (presumably mythical) creatures exist.

Some of the most convincing scientific existence proofs submitted recently concern nanotechnology—a field of engineering that focuses on the manipulation of matter at the molecular level. Nature, it turns out, routinely engages in nanotechnology by shaping proteins into useful configurations and by harvesting the information contained in DNA molecules in order to transform that information into organisms similar to ourselves.

As Seth Shostak, senior astronomer at the SETI Institute, has written about my Selfish Biocosm hypothesis:

> Ever since Newton, scientists have tried to understand existence by discovering its underlying rules. The result has been a massive edifice of natural law, and biology has been seen as a consequence of the universe's construction, rather than an instigator. Only on Earth's surface, where life has molded the seas, the continents, and even the atmosphere, is biology thought to have had an important role in shaping physical conditions—the so-called Gaia hypothesis. But Gardner has taken Gaia to its furthest conceivable magnitude: extending the role and influence of life to the stars and beyond.[16]

Implications of the Selfish Biocosm Hypothesis

The Selfish Biocosm hypothesis places the story of our origin and destiny in a cosmic context.

What is humankind's place in the universe? That fundamental question underlies both scientific inquiry and millennia of religious thought. The traditional answer of science is that life and human intelligence are of no cosmic consequence, but merely the random outcome of the interplay of natural forces. Mainstream religion answers the same question in many different ways, but most share the view that the mind of the Creator of the universe is ultimately inaccessible to mortal minds. The Selfish Biocom hypothesis challenges both viewpoints and suggests that the emergence of life and mind is a cosmic imperative encoded in the basic laws of nature and, further, that highly evolved intelligence will eventually play the key role in reproducing the cosmos.

The Selfish Biocosm hypothesis provides the framework for a new style of final theory.

The hypothesis suggests that in attempting to explain the linkage between life, intelligence, and the bio-friendly qualities of the cosmos, most mainstream scientists have, in essence, been peering through the wrong end of the telescope. The book asserts that life and intelligence are, in fact, the primary cosmological phenomena and that everything else—the constants of inanimate nature, the dimensionality of the universe, the origin of carbon and other elements in the hearts of giant supernovas, the pathway traced by biological evolution—is secondary and derivative. In the words of British Astronomer Royal Sir Martin Rees, the hypothesis embraces the proposition that "what we call the fundamental constants—the numbers that matter to physicists—may be *secondary consequences* of the final theory, rather than direct manifestations of its deepest and most fundamental level."[17] Rees's insight yields a glimpse of a new kind of final theory that views the oddly bio-friendly qualities of our anthropic universe—a universe adapted to the peculiar needs of carbon-based living creatures just as thoroughly as those creatures are adapted to the physical exigencies of the universe—not as an irksome curiosity but rather as a vital set of clues pointing toward a radically new vision of the basic nature of the cosmos. The Selfish Biocosm hypothesis attempts to follow those clues to their logical conclusion.

The hypothesis provides the foundation for a new set of ethical imperatives and insights.

Science should not divorce itself from the ethical, legal, and social implications of new theories. My first book identified three key ethical imperatives and insights that derive from the new cosmological theory:

First, that humankind is ethically obliged to safeguard the welfare of future generations.

Second, that a spirit of species-neutral altruism should inform our interactions with other living creatures and with the environment we share.

Third, that we and other living creatures throughout the cosmos are part of a vast, as yet undiscovered trans-terrestrial community of lives and intelligences spread across billions of galaxies and countless parsecs who are collectively engaged in a portentous mission of truly cosmic importance. Under the biocosm vision, we share a common fate with that community—to help shape the future of the universe and transform it from a collection of lifeless atoms into a vast, transcendent mind.

The inescapable implication of the Selfish Biocosm hypothesis is that the immense saga of biological evolution on Earth is one tiny chapter in an ageless tale of the struggle of the creative force of life against the disintegrative acid of entropy, of emergent order against encroaching chaos, and ultimately of the heroic power of mind against the brute intransigence of lifeless matter. Through the quality and character of our contribution to the progress of life and intelligence in this epic struggle, we shape not only our own lives and those of our immediate progeny, but the lives and minds of every generation of living creatures down to the end of time. We thereby help to shape the ultimate fate of the cosmos itself.

The Selfish Biocosm hypothesis offers a radically novel explanatory paradigm that unites three seemingly disparate aspects of nature—the emergence and proliferation of life, the dawn and accelerating ascendance of intelligence (both natural and artificial), and the apparently mindless evolution of the physical universe—as manifestations of a single universal phenomenon of cosmic ontogeny. The proposed fusion of these three superficially dissimilar and seemingly causally disconnected aspects of nature rests on what NASA historian Steven Dick calls the biological universe worldview: a notion that the universe is primordially biocentric rather than merely mechanistic in nature. In my view, the most plausible and parsimonious extrapolation from this worldview—the easiest way of capturing its stunning implications—is to simply hypothesize that the cosmos is nothing more and nothing less than a vast emerging life: a biocosm.

If one makes the heroic assumption that the Selfish Biocosm hypothesis is correct, a new mystery quickly emerges: How could the process of life-mediated cosmological reproduction that is the heart of the hypothesis have ever gotten started in the first place?

Chapter 9

Alpha ↔ Omega

For the first half of the 20th century, the consensus view among astrophysicists was that the universe existed in an eternal steady state, in which ceaseless expansion of the fabric of space-time was exquisitely and precisely balanced by the constant creation of new matter in the vast emptiness of interstellar space. Although it was recognized early in the century that Einstein's pioneering theory of relativity strongly implied either expansion or contraction of the universe, mainstream theorists (including Einstein) rejected this possibility, essentially on aesthetic grounds.

The notion that the universe began—and presumably would end—in abrupt singularities at opposite ends of time (by means of the Big Bang and Big Crunch) seemed distressingly similar to biblical stories of cosmic creation *ex nihilo* and eventual Armageddon. The opposing view of an eternal, steady-state universe seemed more elegant and beautiful to theorists, but it turned out to be as fragile and vulnerable to observational falsification as Ptolemy's elaborate system of epicyclical astrophysics. Einstein recognized that merely maintaining the stability of a steady-state universe against the attractive force of gravity required the insertion by hand of a cosmological constant term into the equations of general relativity. As Princeton physicists J. Richard Gott, III, and Li-Xin Li have noted, Einstein's motivation for inserting this arbitrary term into his equations appears to have been primarily non-scientific:

> The Einstein static universe appears to be the geometry Einstein found *a priori* most aesthetically appealing, thus presumably he started with this preferred geometry and substituted it into the field equations to determine the energy-momentum tensor required to produce it. He found a source

> term that looks like dust (stars) plus a term that was proportional to the metric which he called the cosmological constant. The cosmological constant, because of its homogenous large negative pressure, exerts a repulsive gravitational effect offsetting the attraction of the stars for each other; allowing a static model which could exist (ignoring instabilities, which he failed to consider) to the infinite past and future. If one did not require a static model, there would be no need for the cosmological constant.[1]

The ultimate failure of this decidedly unscientific stratagem, which Einstein later bemoaned as the biggest mistake of his scientific career (following Hubble's discovery of clear evidence that the universe was expanding), should serve as a cautionary reminder to all theorists who are motivated primarily by a romantic quest for beauty and elegance in the laws of physics. Nature, it turns out, is not always as beautiful or elegant as final-theory dreamers might wish to imagine. Sometimes Mother Nature is downright messy, as well as surpassingly bizarre.

Indeed, the supreme irony regarding Einstein's misguided quest is that something resembling his cosmological constant has now reappeared in astrophysics in the form of mysterious dark energy. Yet far from imbuing state-of-the-art M-theory-based cosmological models with elegance, the mysteriously minute value of experimentally observed dark energy—not quite but almost zero—is now widely acknowledged as perhaps the most serious impediment to promulgation of a mathematically beautiful and invariant final cosmological theory. In the rueful phrase of one physicist, the discovery of the tiny value of dark energy—a value many orders of magnitude smaller than the value that supersymmetry theory predicts—has transformed theoretical cosmology into a messy environmental science, with all the complications of biology!

Einstein's abrupt abandonment of his arbitrarily inserted cosmological constant and the waning interest in steady-state approaches to cosmology following the discovery of seemingly incontrovertible evidence of the Big Bang offers a cautionary lesson in a second and more subtle sense. It may turn out that Fred Hoyle (the most prominent proponent of the steady-state theory during the 20th century) was largely correct about the basic nature of steady-state cosmogenesis, and erred only in underestimating the sheer scale of the hypothesized process. There is an important lesson here, grounded in a fundamental precept of Darwinism: Scientists who speculate about the nature of the universe must be constantly mindful of the blinders and limitations imposed on puny human intellects by our evolutionary heritage as primates. Put simply, our mental proclivities, instincts, and capacities were shaped by the struggle to survive and reproduce in a very specific terrestrial environment. Scientists are accordingly obliged, in the inimitable phrase of John B. S. Haldane, to constantly remind themselves that the universe we inhabit may be not only queerer than we imagine, but more unusual than we can possibly imagine using our evolutionarily developed senses, sensibilities, and mental capacities.

Fred Hoyle: The Very Model of a Scientific Iconoclast

The vast scale of the cosmos confounds our imagination.

What human mind—calibrated by natural selection to appreciate intuitively the dimensions of African savannahs, primeval arboreal hideaways, and Ice Age mammoth hunting grounds—can truly grasp its enormity? Billions of galaxies, each containing hundreds of billions of stars, those stars probably orbited by trillions of planets, and the entire fabric of space-time expanding outward the way the surface of an inflating balloon does—this is the surpassingly strange picture of our universe that constitutes the consensus paradigm of modern cosmology.

That vision is centered on the premise of a Big Bang—a primordial explosion that launched the whole shebang hurtling outward at breakneck speed—which seems, from a commonsense perspective, perfectly outrageous. What came *before* the Big Bang, we wonder? What *caused* this peculiar genesis event? Could the cosmos really have been born *ex nihilo*, for no apparent reason, and from the loins of nothing at all?

These were the puzzles that led a giant of British astronomy—Fred Hoyle—to suggest a dramatic alternative: The steady-state theory, which hypothesized that the universe is eternal and ever-expanding and that the cosmic storehouse of matter is constantly replenished through a process of continuous creation. As Simon Mitton recently demonstrated in a superb new biography of Hoyle, titled *Conflict in the Cosmos*,[2] the great scientist's genius lay in his ability to resist the temptation to surrender to mainstream orthodoxy. Though Hoyle's cosmological theory may have turned out to have been spectacularly wrong, what cannot be denied is that his stubborn unwillingness to bow to conventional wisdom was a valuable intellectual asset that benefited the entire scientific community. As Mitton put it:

> Hoyle's personal contribution to the rebirth of British astronomy came from his outstanding ability to think outside the box.... He always had a deep conviction that in his "search for the truth,"...any opponent should be able to provide a counterargument from experiment or direct observation. He declined all opposition based on semantic arguments...or appeals to common sense.[3]

Iconoclasm and catholicity of scientific interest were the two key markers of Hoyle's long and conflict-laden life. As astrophysicist Owen Gingerich observes in a thoughtful foreword to *Conflict in the Cosmos*, these characteristics—deeply rooted in Hoyle's hard-scrabble background—were both his greatest strength and the source of his ultimate undoing:

> Fred Hoyle was the quintessential outsider, entering Emmanuel College Cambridge from an impoverished family background and with a distinct Yorkshire accent, and leaving Cambridge in a misguided huff 39 years later. But in between he ascended into the highest ranks of British science, almost single-handedly returning Britain to the top echelons of international theoretical astrophysics and setting it on the path toward excellence in observational astronomy.[4]

The 39-year interregnum was the central chapter in the scientist's life—a tumultuous period characterized by heroic accomplishment, intense controversy, and an extraordinary level of celebrity, which Hoyle achieved both as a popular BBC commentator and as a highly successful science fiction writer. As Mitton points out:

> After 1950, Fred Hoyle was a very public figure at home and abroad.... His broadcasts for the BBC in 1950 were just extraordinary and brought him immediate fame as a gifted expositor.[5]

Hoyle is remembered most vividly for the idea about which he was famously mistaken: that the universe exists in a steady state, with the stockpile of atoms in an eternally expanding cosmos continuously refilled by the constant creation of new matter. Normally, falsified scientific hypotheses such as the steady-state conjecture are tossed unceremoniously in the dustbin of intellectual history, serving at best as amusing footnotes to the main body of orthodox theory (think of Darwin's misplaced reliance on Lamarckism as a subsidiary engine of evolution in *The Origin of Species*). But, once again, Hoyle confounds tradition. Because he was both passionate and brutally honest about the implications of his steady-state hypothesis, Hoyle was able to foment a heated intellectual debate that significantly advanced our understanding of the universe, despite the fact that his particular conjecture turned out to be deeply flawed. As Mitton noted, "What is extraordinary about Fred Hoyle's science is that his impact derives equally from when he was right and when he was wrong!"[6]

If Hoyle was wrong about the nature of the process of cosmogenesis, he was spectacularly right about an equally profound mystery: the origin of the chemical elements. In what is surely his most important contribution to astrophysics, Hoyle and three collaborators were able to demonstrate rigorously in their famous B^2FH scientific paper that all of the elements of the periodic table except the lightest are forged in the hearts of giant supernovae, under a variety of physical conditions, through a process known as nucleosynthesis. It is this process of stellar alchemy, Hoyle and his colleagues showed, that accounts for the richness and complexity of the chemical palette of the universe, which in turn accounts for the possibility of life.

In the midst of this monumental accomplishment, Hoyle stumbled across a deep mystery that eventually lured him away from the shoreline of genuine science out onto the trackless sea of metaphysical speculation: the apparent fine-tuning of nature evidenced by the details of the process through which the element carbon is synthesized.

This discovery provoked Hoyle's most controversial conjecture: the notion that the universe appeared to be deliberately fine-tuned to favor the emergence of carbon-based life. As Hoyle wrote late in his life:

> The issue of whether the universe is purposive is an ultimate question that is at the back of everybody's mind.... And Dr. [Ruth Nanda] Ashen has now raised exactly the same question as to whether the universe is a product of thought. And I have to say that that is also my personal opinion, but I can't back it up by too much of precise argument. There are many aspects of the universe where you either have to say there have been monstrous coincidences, which there might have been, or, alternatively, there is a purposive scenario to which the universe conforms.[7]

The debate over this portentous issue rages on to this day, fueled by the recent discovery of the monstrously large landscape of alternate versions of low-energy physics mathematically allowed by M-theory, only a tiny fraction of which would permit the emergence of anything resembling our own universe and of carbon-based life. Indeed, that discovery has led many cutting-edge cosmologists to overlay a refinement of Big Bang inflation theory, called eternal inflation, with an explanatory approach that has been traditionally reviled by most scientists known as the weak anthropic principle. (The weak anthropic principle merely states in tautological fashion that, because human observers inhabit this particular universe, it must perforce be life-friendly or it would not contain any observers resembling ourselves.) Eternal inflation, developed by Russian-born physicist Andrei Linde, asserts that instead of just one Big Bang there are, always have been, and always will be an infinite multiplicity of Big Bangs going off in inaccessible regions all the time. These Big Bangs create a vast horde of new universes constantly, and the whole ensemble constitutes a multiverse.

One gets the uneasy feeling that if this current theorizing turns out to be correct, Fred Hoyle may have been on the right track all along! Perhaps the multiverse *is* eternal. Perhaps there *is* a process of continuous creation (eternal inflation) as opposed to a one-off genesis event (a single Big Bang).

Maybe the only thing Fred Hoyle truly failed to grasp was the sheer, unexpected grandeur of steady-state cosmogenesis. Hoyle believed that the continuous-creation process yielded "no more than one atom in the course of a year in a volume equal to Saint Paul's Cathedral."[8] This is an image of a natural process comfortably within the confines of our biologically evolved human imagination.

But if Linde and his colleagues are correct, the process of continuous creation operates at a scale utterly beyond our capacity to physically envision it—not mere atoms but entire new baby universes are continuously created in an eternal process with striking parallels to Hoyle's discarded steady-state cosmological theory.

An approach to cosmology that embraces the messiness of the string theory landscape, as well as the continuous-creation premises of eternal chaotic inflation, is steadily gaining credibility within the scientific community. This approach does not seek to overthrow the canonical paradigm of the Big Bang, but rather embeds that supposedly unique event in a (perhaps) eternal sequence of Big Bangs that have created and will continue to create, *ad infinitum*, not only our particular cosmos but also universes other than our own.

An intriguing variation on this basic approach—the basic approach being characterized as enthusiastic embrace of the anthropic landscape of string theory coupled with endorsement of the possibility that eternal inflation may provide a mechanism for actually populating the landscape with a wildly diverse population of baby universes, only a tiny fraction of which are bio-friendly—was explored by J. Richard Gott, III and Li-Xin Li of Princeton in a seminal 1997 paper titled "Can the Universe Create Itself?"[9]

Gott and Li began by acknowledging that "[t]he question of first-cause has been troubling to philosophers and scientists for over two thousand years"[10] including the leading intellectual lights of ancient Greece:

> Aristotle found this sufficiently troubling that he proposed avoiding it by having the Universe exist eternally in both the past and future. That way, it was always present and one would not have to ask what caused it to come into being.[11]

In the spirit of Aristotle, Gott and Li audaciously suggest that asking how to create the universe out of nothing might be posing the wrong question. Why? Because a remarkable property of Einstein's theory of general relativity is that it allows solutions that have closed timelike curves, or CTCs—hypothetical configurations of time and space where gravity is sufficiently strong to bend the space-time continuum into a looping configuration that allows future events to influence the past by permitting, in the words of British mathematician Roger Penrose, "a signal to be sent from some event into the past of that same event."[12]

Using sophisticated mathematics, Gott and Li demonstrate that the possibility of CTCs offered an elegant solution to the question of what came before the Big Bang:

> The question of first-cause has been a troubling one for cosmology. Often, this has been solved by postulating a universe that has existed forever in the past. Big Bang models supposed that the first-cause was a singularity, but questions about its almost, but not quite, uniformity remained.... Ultimately, the problem seems to be how to create something out of nothing.[13]

The unconventional alternative proposed by Gott and Li was essentially that the universe might actually be its own mother:

> In this paper, we consider...the notion that the Universe did not arise out of nothing, but rather created itself. One of the remarkable properties of the theory of general relativity is that in principle it allows solutions with CTCs. Why not apply

> this to the problem of the first-cause? Usually the beginning of the universe is viewed like the south pole. Asking what is before that is like asking what is south of the south pole, it is said. But as we have seen, there remain unresolved problems with this model. If instead there were a region of CTCs in the early universe, then asking what was the earliest point in the Universe would be like asking what is the easternmost point on the Earth. There is no easternmost point—you can continue going east around and around the Earth. Every point has points that are east of it. If the Universe contained an early region of CTCs, there would be no first-cause. Every event would have events to its past. And yet the Universe would not have existed eternally in the past. Thus, one of the most remarkable properties of general relativity—the ability in principle to allow CTCs—would be called upon to solve one of the most perplexing problems in cosmology.[14]

This paper interested me intensely, not merely because of its undeniable brilliance and novelty, but because it offered a potential pathway forward with regard to solving a central problem raised by my Selfish Biocosm hypothesis: Who or what initially launched the process of cosmic reproduction? How did life and intelligence first become possible? How did the first universe become sentient and thus capable of seeding its progeny with bio-friendly physical laws and constants? These were questions I was constantly asked in seminars and lectures I gave about my hypothesis. And if the ideas of Gott and Li offered key elements of a potential solution to this conundrum, I was prepared to take their wonderfully bizarre solution seriously.

How Vast Is Our Ignorance, How Queer Is Our Cosmos

The most difficult thing for laypeople to understand about science, Nobel laureate Murray Gell-Mann once told me after a lecture in Portland, Oregon, is how very little scientists truly comprehend about the basic nature of nature, how vast is our ignorance of the fundamental reality of the cosmos. Gell-Mann's statement reminded me of the comment of that supreme master of quantum physics, Nobel laureate Richard Feynman, who remarked famously: "I think it can be safely said that nobody understands quantum mechanics."

Isaac Newton, the father of modern physics and author of the *Principia*—arguably the single most sublime achievement of the human intellect—made exactly the same point some 300 years earlier when he said:

> I do not know what I may appear to the world, but to myself I seem only like a boy playing on the seashore, and diverting myself in now and then finding a smoother pebble or a prettier shell than ordinary, whilst the great ocean of truth lay all undiscovered before me.[15]

Why is it that our greatest geniuses—Gell-Mann, Feynman, Newton, and their ilk—can humbly concede how pitifully limited is the reach of deep human insight and comprehension, while lesser spirits noisily proclaim the certainty of their conclusions and forcefully dismiss any dissent, doubt, or skepticism?

The short answer, I think, is that humans crave certainty, even false certainty, in preference to the sense of vertigo induced by a clear-eyed acknowledgment that we are, at least most of the time, stumbling in the dark down an unmarked path through the baffling wilderness of an unknown—perhaps unknowable—reality. Facing up to the limits of our knowledge and the enormity of our ignorance is an acquired skill, to put it mildly. But it is a skill worth cultivating. For if we don't realize where the shoreline of reasonably well-established scientific theory ends and the vast sea of undiscovered truth begins, how can we possibly hope to measure our progress toward a deeper and more encompassing scientific enlightenment?

That is the *leitmotif* of Oxford physicist Roger Penrose's inappropriately subtitled *The Road to Reality: A Complete Guide to the Laws of the Universe*.[16] The great virtue of this massive tome—an imposing brick of a book that bristles with equations and weighs in at a daunting 1,099 pages—is not so much the encyclopedic review it provides of the history and current status of theoretical physics, but rather its forthright acknowledgment that, for all its magnificent achievements, modern science remains a flickering candle surrounded by inky mystery. With refreshing candor, Penrose forcefully underscores the extent of our ignorance about the origin—and indeed the very nature—of the physical laws that govern the operation of our magnificent universe. What the enterprise of science has compiled to date, in Penrose's view, is a decidedly *incomplete* guide to the laws of nature:

> I hope that it is clear, from the discussion given in the preceding sections that our road to understanding the nature of the real world is still

> a long way from its goal. Perhaps this goal will never be reached, or perhaps there will eventually emerge some ultimate theory, in terms of which what we call "reality" can in principle be understood. If so, the nature of that theory must differ enormously from what we have seen in physical theories so far.[17]

The great contrarians of science—British cosmologist Fred Hoyle, Princeton physicists John Wheeler and Freeman Dyson, Tommy Gold, Francis Crick, Albert Einstein, and Roger Penrose—play a critically important role in maintaining the intellectual integrity of the scientific enterprise. As Gold put it perfectly a few years ago:

> New ideas in science are not right just because they are new. Nor are old ideas wrong just because they are old. A critical attitude is clearly required of every seeker of truth. But one must be *equally* critical of both the old ideas as of the new. Whenever the established ideas are accepted uncritically and conflicting new evidence is brushed aside or not even reported because it does not fit, that particular science is in deep trouble.[18]

Penrose, a quintessential scientific contrarian, is consumed by the fear that uncritical adherence to ideas deemed mainstream by the scientific establishment might blind us to theoretical possibilities that, though radically novel, may offer a deeper understanding of our queer cosmos. He strives heroically in *The Road to Reality* to forestall that intellectual calamity, reminding us in the process of just how enormous is the distance that we have yet to travel down the road that may someday lead to a full understanding of reality.

Maybe the ideas of Gott and Li—that the universe might conceivably be its own mother—will turn out to be the kind of subtle shift in perspective that will allow us to penetrate the enduring mystery of first-cause and cosmic origin.

As I began to overlay the CTC cosmic origin concept of Gott and Li with key elements of my Selfish Biocosm hypothesis, it slowly dawned on me that their ideas might indeed provide the intellectual scaffolding for a satisfactory answer to the tantalizing question of who or what generated the first life-friendly cosmos and endowed it with the capacity to not only nurture life and intelligence, but also to propagate life-friendly successor universes. But the more I fleshed in this picture, the more I began to realize what an utterly bizarre portrait of the cosmos was beginning to emerge!

The first step in my thinking was to remind myself how problematic—indeed, enigmatic—was the role of time in the laws of physics. As Sean Carroll, a cosmologist at the University of Chicago, put it, "The [observed] arrow of time in our universe is puzzling because the fundamental laws of physics themselves are symmetric and don't seem to discriminate between the past and future."[19] Most physicists attribute the perceived future-oriented directionality of time to the effect of entropy: The universe is slowly unwinding from a stunningly low-entropy (highly ordered) state that prevailed at the moment of the Big Bang into the relatively high-entropy (relatively disordered) cosmic state that prevails at present. The process of cosmic transition from low- to high-entropy states, these scientists hypothesize, accounts for the perceived unidirectionality of time.

Yet this formulation seems overly facile. For instance, if our Big Bang was birthed as the result of a random quantum fluctuation in a pre-existing space-time continuum that was itself very old and thus characterized by a very high-entropy state, was the transition to the low-entropy state of our Big Bang tantamount to a reversal in the direction of time as it was flowing in the predecessor continuum? More fundamentally, as Carroll has argued, within a vast multiverse (consisting of the universe created by our Big Bang together with a host of other universes birthed by their own Big Bangs) time may be "actually symmetric, and the laws [of physics] truly don't care about which direction it is moving."[20] As he puts it:

> In our patch of the cosmos, time just so happens to be moving forward because of its initial low entropy, but there are others where this is not the case. The far past and the far future are filled with these other baby universes, and they would each think that the other had its arrow of time backwards. Time's arrow isn't a basic aspect of the universe as a whole, just a hallmark of the little bit we see.[21]

The fact that the perceived flow of time in our neck of the multiverse happens to be exclusively future-oriented, Carroll believes, is just a random quirk of cosmic fate:

> Over a long enough period of time, a baby universe such as ours would have been birthed into existence naturally. Our observable universe and its hundred billion galaxies is just one

> of those things that happens every once in a while, and its arrow of time is just a quirk of chance due to its beginning amid a sea of universes.[22]

If this discussion about the quirky mysteries of time is beginning to make your head spin, better grab a bottle of aspirin, because things are only going to get weirder from here on!

Remember Isaac Newton, the father of modern physics? For Newton, time was platonically pure and mathematically perfect. Its exclusively forward flow from past to future was implacable and invariant in tempo. As he put it in his sublime masterwork *Principia Mathematica*, "Absolute, true, and mathematical Time, of itself, and from its own nature, flows equably without relation to anything external."

So matters rested comfortably until an unknown Swiss patent clerk published, in 1905, his revolutionary special theory of relativity. As Richard Talcott has noted:

> In 1905's special theory of relativity, [Einstein] started with two postulates: The laws of physics should look the same to every observer in uniform motion (moving in a straight line at constant speed), and the speed of light in a vacuum should be the same for every observer in constant motion. The only way these two conditions can be met simultaneously is if time passes at different rates for different observers.[23]

Einstein's special theory of relativity established a simple but utterly counter-intuitive rule: The faster an observer travels, the slower time passes for that observer. This time dilation effect, which only becomes obvious as an observer (or an accelerated sub-atomic particle) nears the speed of light, has been repeatedly verified by a host of ingenious experiments.

When Einstein turned his attention to the effect of gravity on the flow of time in his 1915 general theory of relativity, things got even stranger. Time, it turns out, slows down in the presence of gravity; in the presence of an extremely strong gravitational field such as that surrounding a collapsed neutron star, time would flow at only about 75 percent of the rate at which it flows on Earth. Precise atomic clocks can even measure minute differences in the flow of time on Earth's surface (slightly slower because of the strength of terrestrial gravity) and in satellites orbiting the Earth (slightly faster because of the diminution in the strength of Earth's gravity).

When gravity reaches absurd strength, a black hole is born. At the event horizon of a black hole, time actually comes to a screeching halt. Thus, astonishingly, at the event horizons of the massive black holes that scientists believe lie at the center of most galaxies (including our own Milky Way galaxy), time actually stands still.

But that's not the strangest aspect of what Caltech physicist Kip Thorne has called Einstein's outrageous intellectual legacy. Einstein's theory of general relativity allows solutions that have closed timelike curves or CTCs. These solutions were discovered by the brilliant logician and mathematician Kurt Gödel, who worked alongside Einstein at the Institute for Advanced Studies in Princeton, New Jersey.

The philosophical implications of the potential existence of CTCs are profound indeed: CTCs imply that the future can actually influence and causally reshape the past!

In Gödel's view, the ontological implications were even more radical. As philosopher of science Palle Yourgrau has observed, the existence of CTCs implied to Gödel that "time itself—hence speed and motion—is but an illusion."[24] As Yourgrau put it:

> [I]f we can revisit the past, it still exists. How else could it be revisited? You can't visit New Jersey if New Jersey is no longer there, and you can't return to time *t* if *t* has departed from the realm of existence. Thus temporal distance—past and future—turns out to be as ontologically neutral as the measure of space.... For Gödel, if there is time travel, there isn't time. The goal of the great logician was not to make room in physics for one's favorite episode of *Star Trek*, but rather to demonstrate that if one follows the logic of relativity further even than its father was willing to venture, the results will not just illuminate but eliminate the reality of time.[25]

The second great revolution in 20th-century physics was the birth of quantum mechanics. As with Einstein's relativity revolution, the quantum revolution offers startlingly counterintuitive insights about the very nature of time. For instance, the experimentally tested phenomenon of quantum nonlocality means that even widely separated quantum-entangled particles can somehow transmit signals to one another instantaneously, although the speed of light theoretically imposes a maximum limit on the velocity with which information can be transmitted. Indeed, there is reason to think that a pair of entangled particles located at opposite ends of the visible universe could signal one another instantaneously, even though separated by billions of light-years. Such faster-than-light, or *transluminal*, signaling may, at least under one leading interpretation of quantum phenomena, consist of the transmission of information *backward* in time. As Johnjoe McFadden put it in *Quantum Evolution*:

> The *transactional interpretation* of quantum mechanics...sacrifices the principle of locality by allowing signals to travel backwards in time. This approach grew out of work by Richard Feynman and John Wheeler who suggested that electromagnetic waves may travel both forward and backward in time. John Cranmer of the University of Washington in Seattle extended this suggestion to propose that EPR correlations [between quantum-entangled particles] are established by particles signaling backward-in-time.[26]

Moreover, because many cosmologists now regard the origin of the Big Bang as a quantum event, then theoretically every particle in the cosmos was originally entangled with every other particle. According to McFadden:

> If we go back far enough, right back to the Big Bang, then all particles in the universe have interacted. Every particle becomes connected to every other particle in a single massively entangled super-EPR quantum state. This state will persist until measured. The Copenhagen interpretation is then stark—the world does not exist until we (whoever *we* are—that is not made clear) measure it.[27]

The most extreme statement of this point of view was offered by the great Princeton physicist John Wheeler. Termed variously *it from bit* or the *participatory anthropic principle*, Wheeler's approach represents a radical extrapolation from the observer-focused Copenhagen interpretation of quantum phenomena. Put simply, Wheeler proposes that reality does not exist until it is observed. This implies, among other weird things, that emerging future states of a universe containing conscious observers continuously recreate and reshape past cosmic states. As McFadden puts it:

> How real was the universe before consciousness evolved? The physicist John Wheeler has taken the consciousness-dependent reality view to its logical conclusion, proposing that we live in a "participatory universe," wherein the universe depends for its existence on conscious observers to make it real, not only today but retrospectively right back to the Big Bang! Wheeler suggests that the presence of observers imparts a "tangible 'reality' to the universe, not only now but back to the beginning," by a kind of backward-acting wave function collapse. In this scenario, the universe existed in an undetermined ghost state until the first conscious being opened its eyes to collapse the wave function for the entire universe and bring into being its entire history, including the geological and fossil record recording its own evolution.[28]

Keeping in mind the utter weirdness of the reformulated notions of time offered by both relativity theory and quantum mechanics, I next compared Gott and Li's scenario with an alternative to standard inflationary theory called the modified ekpyrotic cyclic universe scenario. I had dealt with this theory in another scientific paper that I published in the *Journal of the British Interplanetary Society*, which probed the possibility that this scenario might be consistent with a cosmological evolutionary process that would culminate in a cosmic state in which the universe would be transformed into a maximally capable computer. As discussed at length in *Biocosm*, the final state of the cosmos prior to and during a Big Bounce stage under the new cyclic scenario appears to be precisely consistent with Seth Lloyd's scientific description in a landmark *Nature* article of a cosmic computer as powerful as the laws of physics would allow.

A self-creating universe
Image provided by Dejan Vinkovic

Beyond offering a scenario in which the evolution of the physical universe might culminate in the emergence of a cosmic computer as powerful as the laws of physics will allow—Seth Lloyd waggishly calls it the ultimate laptop—the new modified ekpyrotic cyclic universe scenario was important in my quest to uncover a "first cause" of life-friendly cosmos for a different reason. One of the peculiar aspects of the scenario is that it predicted that the cosmos would not end in a true final singularity. Rather, it predicted that while the universe undergoes an endless sequence of cosmic epochs that begin with the universe expanding from a Big Bang and end with the universe contracting to a Big Crunch, the flow of time—and thus the flow of causation—continues smoothly across the Big Crunch/Big Bang era. This implies that the fruits of the final computational state hypothesized in my British Interplanetary Society journal paper could conceivably be transmitted to a successor epoch—in effect, transmitted to a successor universe.

Now I was ready to merge this key hypothesized characteristic of the modified cyclic ekpyrotic scenario—the idea that information and causation could conceivably cross the Big Crunch/Big Bang threshold and thereby seed a new universe (or a new cosmic cycle) with a life-friendly cosmic code—with the conjecture of Gott and Li that the universe could conceivably be its own physical mother. The fusion of these two sets of controversial ideas leads to a startling possibility: What we think of as the distant future could conceivably be the source of the information that caused the laws of physics in the very distant past—specifically, at the instant of the Big Bang, which launched our particular universe—to assume their improbably life-friendly values and configurations.

In short, not only the cosmos, but also what Heinz Pagels calls the cosmic code (the hodgepodge of apparently arbitrary physical laws and constants that prevail in our universe), could, under this hypothesis, conceivably be its own mother. This would be an exceedingly strange universe, in which future events influence past events at least as strongly as the past events influence future events. Indeed, it would be a universe in which the past and future actually coevolve and yield what leading complexity theorist Stuart Kauffman memorably calls "order for free."

Under this scenario, Alpha—the beginning of space-time also known as the Big Bang—is precisely adjacent to Omega (the Big Crunch)—along the unbroken loop of a closed timelike curve or CTC. That CTC represents a vast cosmic cycle that could generate both the life-friendly cosmic code of a mother universe and generate as well the template for producing an endless succession of life-friendly baby universes. I put all this together in what has turned out to be my most controversial scientific paper to date. That paper suggests that Alpha and Omega—past and future—are not separate and discrete states of the cosmos, but coevolving participants in a literally endless process of emergence and recurrence. Is this possibility too strange, too counter-intuitive, and too bizarre to take seriously? Judge for yourself! Read my closing argument to the jury of my scientific peers on the concluding pages of my essay, reproduced with the permission of the journal *Complexity* in Appendix B.

Afterword

Childhood's End and the Unity of Everything

Childhood's End

In *Childhood's End*,[1] science fiction master Arthur C. Clarke tells the story of a future time when all the children of the Earth are transformed into a collective mental being—a kind of incorporeal superorganism being prepared for assimilation into something Clarke called the Overmind. It is a poignant novel because it chronicles a final farewell to all that we hold dear as human beings, especially the dream of a human future embodied in our physical progeny.

Clarke's vision is unduly pessimistic, I believe. Humanity and its progeny species are likely to persist side by side with whatever forms of living beings (natural, artificial, or hybrids of both) may emerge in the centuries and millennia ahead. Why do I believe this? Partly as an article of faith and partly on the basis of a biological analogy.

First, the article of faith.

The Comprehensibility of the Cosmos as Covenant

Albert Einstein was famous not only for his scientific genius, but also for his pithy aphorisms about physics, the universe, and just about everything else. One of my favorite Einstein sayings is this: "The most incomprehensible thing about the world is that it is comprehensible."

Einstein's observation was expanded upon by Eugene Wigner in a famous paper published in *Communications in Pure and Applied Mathematics* titled "The Unreasonable Effectiveness of Mathematics in the Natural Sciences."[2] Wigner's paper concludes "that the enormous usefulness of mathematics in the natural sciences is something bordering on the mysterious and that there is no rational explanation for it."[3] Analyzing example after example of the uncanny capacity of pure mathematics to model physical processes that were only later found to correlate, with extraordinary precision, with the predictions and characteristics of the abstract mathematical models, Wigner began to smell a rat:

> Mathematical concepts turn up in entirely unexpected connections. Moreover, they often permit an unexpectedly close and accurate description of the phenomena in these connections. . . . We are in a position similar to that of a man who was provided with a bunch of keys and who, having to open several doors in succession, always hit on the right key on the first or second trial. He became skeptical concerning the uniqueness of the coordination between keys and doors.[4]

It was, Wigner concluded, "not at all natural that 'laws of nature' exist, much less that man is able to discover them."[5] The second point was echoed by Erwin Schrödinger in his little book *What Is Life?*[6] with respect to the then-nascent science of molecular genetics and biology. Schrödinger marveled that "we [human beings], whose total being is entirely based on a marvelous interplay" of the genetic material of two parent cells that exchange genetic material to form a fertilized egg cell that matures into an adult organism of the same species as its parents, could somehow "possess the power of acquiring considerable knowledge about" the process of heredity and ontogeny. The almost miraculous capacity of the human mind to fathom deep mysteries surrounding the very biological phenomena that gave birth to that inquiring mind itself was, for Schrödinger, a mystery that "may well be beyond human understanding."[7]

Perhaps the key to penetrating the Einstein/Wigner/Schrödinger mystery—the incomprehensible comprehensibility of the cosmos, the unreasonable effectiveness of mathematics in the natural sciences, and the baffling capacity of our biologically evolved mind to fathom the puzzles surrounding its own origin using only the tools that Darwinian evolution has bequeathed to humanity—is to simply take seriously Wigner's cryptic acknowledgment "that the laws of nature must have been formulated in the language of mathematics to be an object for the use of applied mathematics."[8] If the laws themselves *are* mathematical formulations, then the "unreasonable" effectiveness of mathematics in modeling them is not startling, but rather entirely reasonable. The real surprise was noted by Schrödinger: How can it be that our biologically evolved minds could have acquired the capacity to probe deep mysteries such as quantum mechanics and relativity theory that seem utterly alien to the primordial African fitness landscape that shaped human evolution?

In seeking a tentative answer to this question, I would like to reemphasize that, under the Selfish Biocosm hypothesis, the physical laws and constants of inanimate nature have two distinct functions: (1) they establish invariant rules in accordance with which the cosmos is organized; and (2) they function as a kind of deep DNA, prescribing a developmental program of cosmological ontogeny that reliably yields life and ever-greater levels of intelligence.

Might the laws and constants of nature conceivably possess a third level of functionality that I will tentatively characterize as pedagogical? Might the seemingly flawless coherence and mathematical beauty of nature's basic laws serve as a kind of attractive scent, enticing humanity into the investigation of nature's mysteries, much as a butterfly is summoned irresistibly into the heart of a flower by the beauty of its petals and chemical magnetism of its aroma? Reflecting on the sublime beauty of mathematics, Bertrand Russell wrote these memorable lines:

> Mathematics, rightly viewed, possesses not only truth, but supreme beauty, a beauty cold and austere, like that of sculpture, without appeal to any part of our weaker nature, without the gorgeous trapping of painting or music, yet sublimely pure, and capable of a stern perfection such as only the greatest art can show. The true spirit of delight, the exaltation, the sense of being more than Man, which is the touchstone of the highest excellence, is to be found in mathematics as surely as in poetry.[9]

Russell's sentiment, one might cynically observe, corresponds to a butterfly's view of the beauty of a flower. Certainly the flower may appear attractive to the butterfly, but nature has co-opted the insect's sensibilities to advance its own objectives. So, too, our sense of the supreme beauty of mathematics and of the flawless coherence of mathematically formulated physical laws may serve grander cosmic purposes than we currently imagine. Similar to the chemical and visual attraction experienced by the butterfly as it approaches a beckoning flower, our mathematical inquisitiveness and the ardor with which we pursue the ever-receding frontier of scientific knowledge may possess a deeper functionality than we realize.

If so, then I would like to think—perhaps naively—that the very comprehensibility of the cosmos to the evolved human mind—what Darwin aptly called the god-like power of our humble human brains to parse the mathematical language of time and space—is a kind of natural covenant: an unspoken promise (from whom or what I cannot say) that humanity, as the medium through which the laws of nature have engineered their own comprehension, will be privileged to participate in the unfolding pageant of cosmic evolution into the indefinite future.

A Bacterium's Tale

The faith part is fine and dandy, but what hard evidence can I proffer to support my skepticism regarding Clarke's doomsday scenario, and to justify my optimism regarding the probable longevity of humanity in an era dominated by super-smart machines and cyborgs? Alas, not much, but here's the best I can come up with.

I submit, as a falsifiable proposition, the following assertion: In the world of nature, no basic form of life ever really dies. It may mutate or evolve or suffer extinctions of those particular incarnations known as species, but it never truly vanishes from the face of the Earth.

Thus, the ancient life-forms known as Archaea, which include extremophile organisms that love to live near superheated undersea vents and subsist on noxious diets of poisonous (to us) gases and chemicals, continue to share the biosphere with recent arrivals such as reptiles and mammals. Likewise, the eubacteria (or *true* bacteria) continue to happily squirm away in just about every conceivable nook and cranny of the planet, including the intestinal tracts of "higher" species like us. Indeed, as Stephen Jay Gould points out in a *Washington Post* essay titled "Planet of the Bacteria,"[10] during the course of a single human life, the number of E. coli that will inhabit a single person's gut far exceeds the total number of people that now live and have ever lived. Even more startling is the fact that, as Dorion Sagan and Lynn Margulis noted in their book *Garden of Microbial Delights*, "Fully 10 percent of our own dry body weight consists of bacteria, some of which, although they are not a congenital part of our bodies, we can't live without."[11]

Gould had this to say about the enduring dominance of the bacterial mode of life:

> We live now in the "Age of Bacteria." Our planet has always been in the "Age of Bacteria," ever since the first fossils—bacteria, of course—were entombed in rocks more than 3 billion years ago. On any possible, reasonable or fair criterion, bacteria are—and always have been—the dominant forms of life on Earth.[12]

We typically fail to "grasp this most evident of biological facts," Gould believed, "in large measure, as an effect of scale. We are so accustomed to viewing phenomena of our scale—sizes measured in feet and ages in decades—as typical of nature."[13] But the fact of bacterial dominance was clear and indisputable for all with microscopes to see. So too was a deeper lesson: The bacterial mode was not only dominant but essentially indestructible. As Gould put it:

> Let us make a quick bow to the flip side of such long domination—to the future prospects that match such a distinguished and persistent past. Bacteria have occupied life's mode from the very beginning, and I cannot imagine a change of status, even under any conceivable new regime that human ingenuity might someday impose upon our planet. Bacteria exist in such overwhelming number and such unparalleled variety; they live in such a wide range of environments, and work in so many unmatched modes of metabolism. Our shenanigans, nuclear and otherwise, might easily lead to our own destruction in the foreseeable future. We might take most of the large terrestrial vertebrates with us—a few thousand species at most.... I doubt that we could ever substantially touch bacterial diversity. The modal organisms cannot be nuked into oblivion or very much affected by any of our considerable conceivable malfeasances.[14]

In explaining the Selfish Biocosm hypothesis, I often analogize the probable cosmic role of human beings to that played by humble mitochondria in a living eukaryotic cell. As I put it in a 2004 lecture at the Hayden Planetarium in New York:

> Let me conclude by asking whether the Selfish Biocosm hypothesis promotes or demotes the cosmic role of humanity. Have I introduced a new anthropocentrism into the science of cosmology? If so, then you should be suspect on this basis alone of my new approach because, as Sigmund Freud pointed out long ago, new scientific paradigms must meet two distinct criteria to be taken seriously: They must reformulate our vision of physical reality in a novel and plausible way and, equally important, they must advance the Copernican project of demoting human beings from the centerpiece of the universe to the results of natural processes.
>
> At first blush, the Selfish Biocosm hypothesis may appear to be hopelessly anthropocentric. Freeman Dyson once famously proclaimed that the seemingly miraculous coincidences exhibited by the physical laws and constants of inanimate nature—factors that render the universe so strangely life-friendly—indicated to him that "the more I examine the universe and study the details of its architecture, the more evidence I find that the universe in some sense knew we were coming." This strong anthropic perspective may seem uplifting and inspiring at first blush but a careful assessment of the new vision of a bio-friendly universe revealed by the Selfish Biocosm hypothesis yields a far more sobering conclusion.
>
> To regard the pageant of life's origin and evolution on Earth as a minor subroutine in an inconceivably vast ontogenetic process through which the universe prepares itself for replication is scarcely to place humankind at the epicenter of creation. Far from offering an anthropocentric view of the cosmos, the new perspective relegates humanity and its probable progeny species (biological or mechanical) to the functional equivalents of mitochondria—formerly free-living bacteria whose special talents were harnessed in the distant past when they were ingested and then pressed into service as organelles inside eukaryotic cells.[15]

Mitochondria are transmuted bacteria. They share with their bacterial cousins and neighbors the virtues of ubiquity, biochemical virtuosity, and—as a mode of life—*de facto* immortality.

If we and our progeny species are indeed the functional equivalents of mitochondria from a cosmic perspective, then it does not seem unreasonable for humanity (notwithstanding the pessimism of Arthur C. Clarke and *Childhood's End*) to aspire to at least a modest measure of longevity as a species. The sheer durability of the bacterial mode of life serves an existence proof that this hope is at least not frivolous.

The Unity of Everything

The *Complexity* essay with which I began my occasionally perilous but ever exhilarating voyage toward the outer limits of cosmological science—"The Selfish Biocosm: Complexity as Cosmology"[16]—ends with this passage from a lecture delivered by Princeton physicist John Wheeler in 1989 at the Santa Fe Institute:

> A single question animates this report: Can we ever expect to understand existence? Clues we have, and work to do, to make headway on that issue. Surely someday, we can believe, we will grasp the central idea of it all as so simple, so beautiful, so compelling that we will all say to each other, "Oh, how could it have been otherwise? How could we all have been so blind so long?"[17]

It is the assertion of this book that "the central idea of it all" is the underlying unity of everything and everyone in the cosmos. As stated in the introduction, the book is the story of an idea, and the idea is quite simple. It is that the best way to think about life, intelligence, and the universe is that they are not separate things but different aspects of a single phenomenon.

We can assign many shorthand names to that universal phenomenon—the biological universe (Steven Dick), vital dust (Christian de Duve), or biocosm—but the sheer wonder of the phenomenon was best captured in the poetic words of Paul Davies:

> [The proponents of a biological universe worldview are making] a huge and profound assumption about the nature of nature. They are saying, in effect, that the laws of the universe are cunningly contrived to coax life into being against the raw odds; that the mathematical principles of physics, in their elegant simplicity, somehow know in advance about life and its vast complexity. If life follows from [primordial] soup with causal dependability, the laws of nature encode a hidden subtext,

> a cosmic imperative, which tells them: "Make life!" And, through life, its by-products: mind, knowledge, understanding. It means that the laws of the universe have engineered their own comprehension. This is a breathtaking vision of nature, magnificent and uplifting in its majestic sweep. I hope it is correct. It would be wonderful if it were correct. But if it is, it represents a shift in the scientific world-view as profound as that initiated by Copernicus and Darwin put together.[18]

Davies' last sentence bears repeating for emphasis: *if this vision of nature is correct, it represents a shift in the scientific world-view as profound as that initiated by Copernicus and Darwin put together.*

Richard Dawkins concluded his monumental masterwork *The Ancestor's Tale* by reflecting on the sheer wonder of the emergence of life and the stunning accomplishments of the evolutionary process in a once-lifeless universe:

> [A]s . . . I reflect on the whole pilgrimage of [evolution], my overwhelming reaction is one of amazement. Amazement not only at the extravaganza of details that we have seen; amazement, too, at the very fact that there are any such details to be had at all, on any planet. The universe could so easily have remained lifeless and simple—just physics and chemistry, just the scattered dust of the cosmic explosion that gave birth to time and space. The fact that it did not—the fact that life evolved out of nearly nothing, some 10 billion years after the universe evolved out of literally nothing—is a fact so staggering that I would be mad to attempt words to do it justice. And even that is not the end of the matter. Not only did evolution happen: it eventually led to beings capable of comprehending the process, and even of comprehending the process by which they comprehend it.[19]

I, too, am amazed and awed by the unified process of cosmic and biological evolution. But I differ from Dawkins with regard to one essential particular. Under my Selfish Biocosm hypothesis, the emergence of life and the evolution of intelligence is literally pre-programmed by the laws and constants of physics, which function similar to cosmic DNA. Contrary to Dawkins's assertion, the universe, in my view, could not have easily remained simple and barren. The emergence of life and intelligence was written into the cosmic playbook from the very first moments of the Big Bang. And life was destined, from that very instant, to eventually dominate the cosmos, infuse it with massive intelligence, and ultimately serve as the instrument of cosmic replication.

The Hubble Ultra-Deep Field Images as Baby Photos

A couple of years ago I was asked by *What Is Enlightment?* magazine to reflect on my impressions of the incredibly distant astronomical objects captured in Hubble Ultra-Deep Field images. Here is what I said:

> In contemplating the image revealed by the Hubble Ultra-Deep Field, I am reminded of the sense of wonder felt by Antoni van Leeuwenhoek, the father of observational microbiology, when he first peered through a primitive microscope and glimpsed vast hordes of 'wee beasties' populating drops of pond water and human spittle. The hidden living firmament that so astonished the Dutch scientist (bacteria, protists, rotifers, nematodes, and much more) turned out to be the very foundation of the global ecosystem—the microscopic cells of Gaia's flesh and blood. So, too, the unsuspected celestial grandeur revealed by this image may someday be appreciated as a poignant baby photo—the faint image of a moment, unfathomably distant in time and space, when the vast universe began an utterly mysterious process of coming to life.[20]

As I write the concluding words of this book, I am sitting on my deck on a warm and sunny September afternoon in Portland, Oregon, overlooking the beautiful Willamette River. And I am thinking of an earlier book, also authored by Richard Dawkins, entitled *River Out of Eden*,[21] that helped shape the ideas that I ended up presenting in that first *Complexity* essay which launched what some will undoubtedly regard as my quixotic cosmological quest.

Dawkins' arresting vision of a hierarchical structure of replicators was the conceptual foundation for the Selfish Biocosm hypothesis. I parted company with Dawkins only with respect to how high and far that hierarchy might eventually reach. As I wrote in the essay:

> The final replication threshold foreseen by Dawkins is Threshold 10 (the "Space Travel Threshold") which he describes as follows:

> After radio waves, the only further step we have imagined in the outward progress of our own explosion is physical space travel itself: Threshold 10, the Space Travel Threshold. Science-fiction writers have dreamed of the interstellar proliferation of daughter colonies of humans, or their robotic creations. These daughter colonies could be seen as seedlings, or infections, of new pockets of self-replicating information–pockets that may subsequently themselves expand explosively outward again, in satellite replication bombs, broadcasting both genes and memes. If this vision is ever realized, it is perhaps not too irreverent to imagine some future Christopher Marlowe reverting to the imagery of the digital river: "See, see, where life's flood streams in the firmament!"

> The momentous question posed [in this essay] can be restated in terms of Dawkins' classification scheme: Is Threshold 10 truly the final replication threshold? Or might there be a Threshold 11, which we may provisionally call the Cosmic Replication Threshold? Might Threshold 11 harbor a radically new type of replicator–differing from the preceding classes as profoundly as the meme differs from the gene but incorporating the complex interactions of those precedent entities as subroutines–which we might provisionally label (in deference to Dawkins' memorable nomenclature) as the Selfish Biocosm replicator class?[22]

This book is the story of Replicator Threshold 11 and what the very existence of such a threshold implies about the nature of the universe, its ultimate fate, and the cosmic role of life and intelligence.

As I gaze out on the shimmering Willamette and reflect on Dawkins's book, I remind myself that *The Intelligent Universe* is also the story of a river—what Dawkins called the river out of Eden. The story goes like this: Beneath the forests and fertile farms and soaring peaks of my verdant Oregon wriggle numberless hordes of those "wee beasties" that van Leeuwenhoek first glimpsed through the lens of his primitive microscope—the colonies of bacteria that form the sturdy foundation of life on Earth. Beneath and within the bodies of those wee beasties—and beneath and within everything else on Earth and every planet and star in the firmament above—are the elements of chemistry's periodic table. These elements began their lives in titanic supernovae explosions millions of light-years away and completed epic journeys to our solar system to form our world and endow it with a rich store of the raw materials needed by living creatures. Beneath the chemical elements are the furiously vibrating snippets of energy known as superstrings, writhing away in a tiny and mysteriously life-friendly corner of the vast 11-dimensional M-theory landscape that we shall perhaps never completely explore or understand. Above this hierarchy floats the elegant grand dame of the whole shebang—our beautiful and perplexing cosmos—that was born from the loins of nothing at all and is waltzing inexorably toward a distant rendezvous with highly evolved life and intelligence, perhaps including our own progeny. And through it all, from Big Bang to Big Crunch to new Big Bang, from Alpha to Omega and back to Alpha, runs a great unstoppable river—an everlasting cosmic flood tide of counter-entropic energy that complexity theorist Stuart Kauffman memorably called the force of anti-chaos. That river, that tide, that force—is life itself.

The Intelligent Universe

Appendix A

The Physical Constants as Biosignature: An Anthropic Retrodiction of the Selfish Biocosm Hypothesis

by James N. Gardner

As published in March 2005, in Issue 3/3
of the *International Journal of Astrobiology,*
Cambridge University Press
reprinted with permission

Abstract

Goal 7 of the NASA Astrobiology Roadmap states: "Determine how to recognize signatures of life on other worlds and on early Earth. Identify biosignatures that can reveal and characterize past or present life in ancient samples from Earth, extraterrestrial samples measured *in situ*, samples returned to Earth, remotely measured planetary atmospheres and surfaces, and other cosmic phenomena." The cryptic reference to "other cosmic phenomena" would appear to be broad enough to include the possible identification of biosignatures embedded in the dimensionless constants of physics. The existence of such a set of biosignatures—a life-friendly suite of physical constants—is a retrodiction of the Selfish Biocosm (SB) hypothesis. This hypothesis offers an alternative to the weak anthropic explanation of our indisputably life-friendly cosmos favored by (1) an emerging alliance of M-theory-inspired cosmologists and advocates of eternal inflation like Linde and Weinberg, and (2) supporters of the quantum theory-inspired sum-over-histories cosmological model offered by Hartle and Hawking. According to the SB hypothesis, the laws and constants of physics function as the cosmic equivalent of DNA, guiding a cosmologically extended evolutionary process and providing a blueprint for the replication of new life-friendly progeny universes.

1. Introduction

The notion that we inhabit a universe whose laws and physical constants are fine-tuned in such a way as to make it hospitable to carbon-based life is an old idea.[1] The so-called "anthropic" principle comes in at least four principal versions[2] that represent fundamentally different ontological perspectives. For instance, the "weak anthropic principle" is merely a tautological statement that since we happen to inhabit this particular cosmos it must perforce by life-friendly or else we would not be here to observe it. As Vilenkin put it recently, "the 'anthropic' principle, as stated above, hardly deserves to be called a principle: it is trivially true."[3] By contrast, the "participatory anthropic principle" articulated by Wheeler and dubbed "it from bit"[4] is a radical extrapolation from the Copenhagen interpretation of quantum physics and a profoundly counterintuitive assertion that the very act of observing the universe summons it into existence.

All anthropic cosmological interpretations share a common theme: a recognition that key constants of physics (as well as other physical aspects of our cosmos such as its dimensionality) appear to exhibit a mysterious fine-tuning that optimizes their collective bio-friendliness. Rees noted[5] that virtually every aspect of the evolution of the universe—from the birth of galaxies to the origin of life on Earth—is sensitively dependent on the precise values of seemingly arbitrary constants of nature like the strength of gravity, the number of extended spatial dimensions in our universe (three of the ten posited by M-theory), and the initial expansion speed of the cosmos following the Big Bang. If any of these physical constants had been even slightly different, life as we know it would have been impossible:

> The [cosmological] picture that emerges—a map in time as well as in space—is not what most of us expected. It offers a new perspective on a how a single "genesis event" created billions of galaxies, black holes, stars and planets, and how atoms have been assembled—here on Earth, and perhaps on other worlds—into living beings intricate enough to ponder their origins. There are deep connections between stars and atoms, between the cosmos and the microworld.... Our emergence and survival depend on very special "tuning" of the cosmos—a cosmos that may be even vaster than the universe that we can actually see.

As stated recently by Smolin,[6] the challenge is to provide a genuinely scientific explanation for what he terms the "anthropic observation":

> **The anthropic observation:** Our universe is much more complex than most universes with the same laws but different values of the parameters of those laws. In particular, it has a complex astrophysics, including galaxies and long lived stars, and a complex chemistry, including carbon chemistry. These necessary conditions for life are present in our universe as a consequence of the complexity which is made possible by the special values of the parameters.

There is good evidence that the anthropic observation is true. Why it is true is a puzzle that science must solve.

It is a daunting puzzle indeed. The strangely (and apparently arbitrarily) biophilic quality of the physical laws and constants poses, in Greene's view, the deepest question in all of science.[7] In the words of Davies,[8] it represents "the biggest of the Big Questions: why is the universe bio-friendly?"

2. Modern History of Anthropic Reasoning

Modern statements of the cosmological anthropic principle date from the publication of a landmark book by Henderson in 1913 entitled *The Fitness of the Environment.*[9] Henderson's book was an extended reflection on the curious fact that there are particular substances present in the environment—preeminently water—whose peculiar qualities rendered the environment almost preternaturally suitable for the origin, maintenance, and evolution of organic life. Indeed, the strangely life-friendly qualities of these materials led Henderson to the view that "we were obliged to regard this collocation of properties in some intelligible sense a preparation for the process of planetary evolution.... Therefore the properties of the elements must for the present be regarded as possessing a teleological character."

Thoroughly modern in outlook, Henderson dismissed this apparent evidence that inanimate nature exhibited a teleological character as indicative of divine design or purpose. Indeed, he rejected the notion that nature's seemingly teleological quality was in any way inconsistent with Darwin's theory of evolution through natural selection. On the contrary, he viewed the bio-friendly character of the inanimate natural environment as essential to the optimal operation of the evolutionary forces in the biosphere. Absent the substrate of a superbly "fit" inanimate environment, Henderson contended, Darwinian evolution could never have achieved what it has in terms of species multiplication and diversification.

The mystery of *why* the physical qualities of the inanimate universe happened to be so oddly conducive to life and biological evolution remained just that for Henderson—an impenetrable mystery. The best he could do to solve the puzzle was to speculate that the laws of chemistry were somehow fine-tuned in advance by some unknown cosmic evolutionary mechanism to meet the future needs of a living biosphere:

> The properties of matter and the course of cosmic evolution are now seen to be intimately related to the structure of the living being and to its activities; they become, therefore, far more important in biology than has previously been suspected. For the whole evolutionary process, both cosmic and organic, is one, and the biologist may now rightly regard the Universe in its very essence as biocentric.

Henderson's iconoclastic vision was far ahead of its time. His potentially revolutionary book was largely ignored by his contemporaries or dismissed as a mere tautology. *Of course* there should be a close match-up between the physical requirements of life and the physical world that life inhabits, contemporary skeptics pointed out, since life evolved to survive the very challenges presented by that pre-organic world and to take advantage of the biochemical opportunities it offered.

While lacking broad influence at the time, Henderson's pioneering vision proved to be the precursor to modern formulations of the cosmological anthropic principle. One of the first such formulations was offered by British astronomer Fred Hoyle. A storied chapter in the history of the principle is the oft-told tale of Hoyle's prediction of the details of the triple-alpha process.[10] This prediction, which seems to qualify as the first falsifiable implication to flow from an anthropic hypothesis, involves the details of the process by which the element carbon (widely viewed as the essential element of abiotic precursor polymers capable of autocatalyzing the emergence of living entities) emerges through stellar nucleosynthesis. As noted by Livio:

> Carbon features in most anthropic arguments. In particular, it is often argued that the existence of an excited state of the carbon nucleus is a manifestation of fine-tuning of the constants of nature that allowed for the appearance of carbon-based life. Carbon is formed through the triple-alpha process in two steps. In the first, two alpha particles form an unstable (lifetime ~ 10^{-16} s)^{8}Be. In the second, a third alpha particle is captured, via ^{8}Be(α, γ)^{12}C. Hoyle argued than in order for the 3α reaction to proceed at a rate sufficient to produce the observed cosmic carbon, a resonant level must exist in ^{12}C, a few hundred keV about the ^{8}Be + ^{4}He threshold. Such a level was indeed found experimentally.[11]

Other chapters in the modern history of the anthropic principle are treated comprehensively by Barrow and Tipler[12] and will not be revisited here.

3. The New Urgency of Anthropic Investigation

Two recent developments have imparted a renewed sense of urgency to investigations of the anthropic qualities of our cosmos. The first is the discovery that the value of dark energy density is exceedingly small but not quite zero—an apparent happenstance, unpredictable from first principles, with profound implications for the bio-friendly quality of our universe. As noted recently by Goldsmith:

> A relatively straightforward calculation [based on established principles of theoretical physics] does yield a theoretical value for the cosmological constant, but that value is greater than the measured one by a factor of about 10^{120}—probably the largest discrepancy between theory and observation science has ever had to bear.

> If the cosmological constant had a smaller value than that suggested by recent observations, it would cause no trouble (just as one would expect, remembering the happy days when the constant was thought to be zero). But if the constant were a few times larger than it is now, the universe would have expanded so rapidly that galaxies could not have endured for the billions of years necessary to bring forth complex forms of life.[13]

The second development is the realization that M-theory—arguably the most promising contemporary candidate for a theory capable of yielding a deep synthesis of relativity and quantum physics—permits, in Bjorken's phrase, "a variety of string vacuua, with different standard-model properties."[14]

M-theorists had initially hoped that their new paradigm would be "brittle" in the sense of yielding a single mathematically unavoidable solution that uniquely explained the seemingly arbitrary parameters of the Standard Model. As Susskind has put it:

> The world-view shared by most physicists is that the laws of nature are uniquely described by some special action principle that completely determines the vacuum, the spectrum of elementary particles, the forces and the symmetries. Experience with quantum electrodynamics and quantum chromodynamics suggests a world with a small number of parameters and a unique ground state. For the most part, string theorists bought into this paradigm. At first it was hoped that string theory would be unique and explain the various parameters that quantum field theory left unexplained.[15]

This hope has been dashed by the recent discovery that the number of different solutions permitted by M-theory (which correspond to different values of Standard Model parameters) is, in Susskind's words, "astronomical, measured not in millions or billions but in googles or googleplexes." This development seems to deprive our most promising new theory of fundamental physics of the power to uniquely predict the emergence of anything remotely resembling our universe. As Susskind puts it, the picture of the universe that is emerging from the deep mathematical recesses of M-theory is not an "elegant universe" but rather a Rube Goldberg device, cobbled together by some unknown process in a supremely improbable manner that just happens to render the whole ensemble fit for life. In the words of University of California theoretical physicist Steve Giddings, "No longer can we follow the dream of discovering the unique equations that predict everything we see, and writing them on a single page. Predicting the constants of nature becomes a messy environmental problem. It has the complications of biology."[16]

4. Two Contemporary Restatements of the Weak Anthropic Principle: Eternal Inflation Plus M-Theory and Many-Worlds Quantum Cosmology

There have been two principal approaches to the task of enlisting the weak anthropic principle to explain the mysteriously small (and thus bio-friendly) value of the density of dark energy and the apparent happenstance by which our bio-friendly universe was selected from the enormously large "landscape" of possible solutions permitted by M-theory, only a tiny fraction of which correspond to anything resembling the Standard Model prevalent in our cosmos.

4.1 Eternal Inflation Meets M-Theory

The first approach, favored by Susskind,[17] Linde,[18] Weinberg,[19] and Vilenkin[10] among others, overlays the model of eternal inflation with the key assumption that M-theory-permitted solutions (corresponding to different values of Standard Model parameters) and dark energy density values will vary randomly from bubble universe to bubble universe within an eternally expanding ensemble variously termed a *multiverse* or a *meta-universe*. Generating a life-friendly cosmos is simply a matter of randomly reshuffling the set of permissible parameters and values a sufficient number of times until a particular Big Bang yields, against odds of perhaps a googleplex-to-one, a permutation that just happens to possess the right mix of Standard Model parameters to be bio-friendly.

4.2 Sum-Over-Histories Quantum Cosmological Model

The second approach invokes a quantum theory-derived sum-over-histories cosmological model inspired by Everett's "many worlds" interpretation of quantum physics. This approach, which has been prominently embraced by Hawking (Hawking and Hertog, 2002), was summarized as follows by Hogan:

> In the original formulation of quantum mechanics, it was said that an observation collapsed a wavefunction to one of the eigenstates of the observed quantity. The modern view is that the cosmic wavefunction never collapses, but only appears to collapse from the point of view of observers who are part of the wavefunction. When Schrödinger's cat lives or dies, the branch of the wavefunction with the dead cat also contains observers who are dealing with a dead cat, and the branch with the live cat also contains observers who are petting a live one.
>
> Although this is sometimes called the "Many Worlds" interpretation of quantum mechanics, it is really about having

> just one world, one wavefunction, obeying the Schrödinger equation: the wavefunction evolves linearly from one time to the next based on its previous state.
>
> Anthropic selection in this sense is built into physics at the most basic level of quantum mechanics. Selection of a wavefunction branch is what drives us into circumstances in which we thrive. Viewed from a disinterested perspective outside the universe, it looks like living beings swim like salmon up their favorite branches of the wavefunction, chasing their favorite places.[21]

Hawking and Hertog have explicitly characterized this "top down" cosmological model as a restatement of the weak anthropic principle:

> We have argued that because our universe has a quantum origin, one must adopt a top down approach to the problem of initial conditions in cosmology, in which histories that contribute to the path integral, depend on the observable being measured. There is an amplitude for empty flat space, but it is not of much significance. Similarly, the other bubbles in an eternally inflating space-time are irrelevant. They are to the future of our past light cone, so they don't contribute to the action for observables and should be excised by Ockham's razor. *Therefore, the top down approach is a mathematical formulation of the weak anthropic principle.* Instead of starting with a universe and asking what a typical observer would see, one specifies the amplitude of interest.[22]

5. Critique of Contemporary Restatements of the Weak Anthropic Principle

Apart from the objections on the part of those who oppose in principle any use of the anthropic principle in cosmology, there are at least three reasons why both the Hawking/Hogan and the Susskind/Linde/Weinberg restatements of the weak anthropic principle are objectionable.

First, both approaches appear to be resistant (at the very least) to experimental testing. Universes spawned by Big Bangs other than our own are inaccessible from our own universe, at least with the experimental techniques currently available to science. So too are quantum wavefunction branches that we cannot, in principle, observe. Accordingly, both approaches appear to be untestable—perhaps untestable in principle. For this reason, Smolin recently argued "not only is the Anthropic Principle not science, its role may be negative. To the extent that the Anthropic Principle is espoused to justify continued interest in unfalsifiable theories, it may play a destructive role in the progress of science."[23]

Second, both approaches violate the mediocrity principle. The mediocrity principle, a mainstay of scientific theorizing since Copernicus, is a statistically based rule of thumb that, absent contrary evidence, a particular sample (Earth, for instance, or our particular universe) should be assumed to be a typical example of the ensemble of which it is a part. The Susskind/Linde/Weinberg approach, in particular, flouts this principle. Their approach simply takes refuge in a brute, unfathomable mystery—the conjectured lucky roll of the dice in a crap game of eternal inflation—and declines to probe seriously into the possibility of a naturalistic cosmic evolutionary process that has the capacity to yield a life-friendly set of physical laws and constants on a nonrandom basis.

Third, both approaches extravagantly inflate the probabilistic resources required to explain the phenomenon of a life-friendly cosmos. (Think of a googleplex of monkeys typing away randomly until one of them, by pure chance, accidentally composes a set of equations that correspond to the Standard Model.) This should be a hint that something fundamental is being overlooked and that there may exist an unknown natural process, perhaps functionally akin in some manner to terrestrial evolution, capable of effecting the emergence and prolongation of physical states of nature that are, in the abstract, vanishingly improbable.

6. The Darwinian Precedent

Hogan has analogized the quantum theory-inspired sum-over-histories version of the weak anthropic principle to Darwinian theory:

> This blending of empirical cosmology and fundamental physics is reminiscent of our Darwinian understanding of the tree of life. The double helix, the four-base codon alphabet and the triplet genetic code for amino acids, any particular gene for a protein in a particular organism—all are frozen accidents of evolutionary history. It is futile to try to understand or explain these aspects of life, or indeed any relationships in biology, without referring to the way the history of life unfolded. In the same way that (in Dobzhansky's phrase), "nothing in biology makes sense except in the light of evolution," physics in these models only makes sense in the light of cosmology.[24]

Ironically, Hogan misses the key point that neither the branching wavefunction nor the eternal inflation-plus-M-theory versions of the weak anthropic principle hypothesize the existence of anything corresponding to the main action principle of Darwin's theory: natural selection. Both restatements of the weak anthropic principle are analogous, not to Darwin's approach, but rather to a mythical alternative history in which Darwin, contemplating the storied tangled bank (the arresting visual image with which he concludes *The Origin of Species*), had confessed not a magnificent obsession with gaining an understanding of the mysterious natural processes that had yielded "endless forms most beautiful and most wonderful," but rather a smug satisfaction that of course the earthly biosphere must have somehow evolved in a just-so manner mysteriously friendly to humans and other currently living species, or else Darwin and other humans would not be around to contemplate it.

Indeed, the situation that confronts cosmologists today is reminiscent of that which faced biologists before Darwin propounded his revolutionary theory of evolution through natural selection. Darwin confronted the seemingly miraculous phenomenon of a fine-tuned natural order in which every creature and plant appeared to occupy a unique and well-designed niche. Refusing to surrender to the brute mystery posed by the appearance of nature's design, Darwin masterfully deployed the art of metaphor[25] to elucidate a radical hypothesis—the origin of species through natural selection—that explained the apparent miracle as a natural phenomenon.

A significant lesson drawn from Darwin's experience is important to note at this point. Answering the question of why the most eminent geologists and naturalists had, until shortly before publication of *The Origin of Species*, disbelieved in the mutability of species, Darwin responded that this false conclusion was "almost inevitable as long as the history of the world was thought to be of short duration." It was geologist Charles Lyell's speculations on the immense age of Earth that provided the essential conceptual framework for Darwin's new theory. Lyell's vastly expanded stretch of geological time provided an ample temporal arena in which the forces of natural selection could sculpt and reshape the species of Earth and achieve nearly limitless variation.

The central point for purposes of this paper is that collateral advances in sciences seemingly far removed from cosmology (complexity theory and evolutionary theory among them) can help dissipate the intellectual limitations imposed by common sense and naïve human intuition. And, in an uncanny reprise of the Lyell/Darwin intellectual synergy, it is a realization of the vastness of time and history that gives rise to the novel theoretical possibility to be discussed subsequently. Only in this instance, it is the vastness of future time and future history that is of crucial importance. In particular, sharp attention must be paid to the key conclusion of Wheeler: most of the time available for life and intelligence to achieve their ultimate capabilities lie in the distant cosmic future, not in the cosmic past. As Tipler has stated, "Almost all of space and time lies in the future. By focusing attention only on the past and present, science has ignored almost all of reality. Since the domain of scientific study is the whole of reality, it is about time science decided to study the future evolution of the universe."[26] The next section of this paper describes an attempt to heed these admonitions.

7. The Selfish Biocosm Hypothesis

In a paper published in *Complexity*,[27] I first advanced the hypothesis that the anthropic qualities which our universe exhibits might be explained as incidental consequences of a cosmic replication cycle in which the emergence of a cosmologically extended biosphere could conceivably supply two of the logically essential elements of self-replication identified by von Neumann: a controller and a duplicating device.[28] The hypothesis proposed in that paper was an attempt to extend and refine Smolin's conjecture[29] that the majority of the anthropic qualities of the universe can be explained as incidental consequences of a process of cosmological replication and natural selection (CNS) whose utility function is black hole maximization. Smolin's conjecture differs crucially from the concept of eternal inflation advanced by Linde[30] in that it proposes a cosmological evolutionary process with a specific and discernible utility function—black hole maximization. It is this aspect of Smolin's conjecture rather than the specific utility function he advocates that renders his theoretical approach genuinely novel.

As demonstrated previously,[31] Smolin's conjecture suffers from two evident defects: (1) the fundamental physical laws and constants do not, in fact, appear to be fine-tuned to favor black hole maximization and (2) no mechanism is proposed corresponding to two logically required elements of any von Neumann self-replicating automaton: a controller and a duplicator.[32] The latter are essential elements of any replicator system capable of Darwinian evolution, as noted by Dawkins in a critique of Smolin's conjecture:

> Note that any Darwinian theory depends on the prior existence of the strong phenomenon of heredity. There have to be self-replicating entities (in a population of such entities) that spawn daughter entities more like themselves than the general population.[33]

Theories of cosmological eschatology previously articulated[34] predict that the ongoing process of biological and technological evolution is sufficiently robust and unbounded that, in the far distant future, a cosmologically extended biosphere could conceivably exert a global influence on the physical state of the cosmos. A related set of insights from complexity theory[35] indicates that the process of emergence resulting from such evolution is essentially unbounded.

A synthesis of these two sets of insights yielded the two key elements of the Selfish Biocosm (SB) hypothesis. The essence of that synthesis is that the ongoing process of biological and technological evolution and emergence could conceivably function as a von Neumann controller and that a cosmologically extended biosphere could, in the very distant future, function as a von Neumann duplicator in a hypothesized process of cosmological replication.

In a paper published in *Acta Astronautica*[36] I suggested that a falsifiable implication of the SB hypothesis is that the process of the progression of the cosmos through critical epigenetic thresholds in its life cycle, while perhaps not strictly inevitable, is relatively robust. One such critical threshold is the emergence of human-level and higher intelligence, which is essential to the eventual scaling up of biological and technological processes to the stage at which those processes could conceivably exert a global influence on the state of the cosmos. Four specific tests of the robustness of the emergence of human-level and higher intelligence were proposed.

In a subsequent paper published in the *Journal of the British Interplanetary Society*[37] I proposed that an additional falsifiable implication of the SB hypothesis is that there exists a plausible final state of the cosmos that exhibits maximal computational potential. This predicted final state appeared to be consistent with both the modified ekpyrotic cyclic universe scenario[38] and with Lloyd's description[39] of the physical attributes of the ultimate computational device: a computer as powerful as the laws of physics will allow.

8. Key Retrodiction of the SB Hypothesis: A Life-Friendly Cosmos

The central assertions of the SB hypothesis are: (1) that highly evolved life and intelligence play an essential role in a hypothesized process of cosmic replication and (2) that the peculiarly life-friendly laws and physical constants that prevail in our universe—an extraordinarily improbable ensemble that Pagels dubbed the cosmic code[40]—play a cosmological role functionally equivalent to that of DNA in an earthly organism: they provide a recipe for cosmic ontogeny and a blueprint for cosmic reproduction. Thus, a key retrodiction of the SB hypothesis is that the suite of physical laws and constants that prevail in our cosmos will, in fact, be life-friendly. Moreover—and alone among the various cosmological scenarios offered to explain the phenomenon of a bio-friendly universe—the SB hypothesis implies that this suite of laws and constants comprise a robust program that will reliably generate life and advanced intelligence just as the DNA of a particular species constitutes a robust program that will reliably generate individual organisms that are members of that particular species. Indeed, because the hypothesis asserts that sufficiently evolved intelligent life serves as a von Neumann duplicator in a putative process of cosmological replication, the biophilic quality of the suite emerges as a retrodicted biosignature of the putative duplicator and duplication process within the meaning of Goal 7 of the NASA Astrobiology Roadmap, which provides in pertinent part:

> Determine how to recognize signatures of life on other worlds and on early Earth. Identify biosignatures that can reveal and characterize past or present life in ancient samples from Earth, extraterrestrial samples measured *in situ*, samples returned to Earth, remotely measured planetary atmospheres and surfaces, *and other cosmic phenomena.*

Does this retrodiction qualify as a valid scientific test of the validity of the SB hypothesis? I propose that it may, provided two additional qualifying criteria are satisfied:

- The underlying hypothesis must enjoy consilience[41] with mainstream scientific paradigms and conjectural frameworks (in particular, complexity theory, evolutionary theory, M-theory, and theoretically acceptable conjectures by mainstream cosmologists concerning the feasibility, at least in principle, of "baby universe" fabrication); and,
- The retrodiction must be augmented by falsifiable predictions of phenomena implied by the SB hypothesis but not yet observed.

9. Retrodiction as a Tool for Testing Scientific Hypotheses

There is a lively literature debating the propriety of employing retrodiction as a tool for testing scientific hypotheses.[42] Oldershaw has discussed the use of falsifiable *retrodiction* (as opposed to falsifiable *prediction*) as a tool of scientific investigation:

> A second type of prediction is actually not a prediction at all, but rather a "retrodiction." For example, the anomalous advance of the perihelion of Mercury had been a tiny thorn in the side of Newtonian gravitation long before general relativity came upon the scene. Einstein found that his theory correctly "predicted," actually retrodicted, the numerical value of the perihelion advance. The explanation of the unexpected result of the Michelson-Morley experiment (constancy of the velocity of light) in terms of special relativity is another example.[43]

As he went on to note, "Retrodictions usually represent falsification tests; the theory is probably wrong if it fails the test, but should not necessarily be considered right if it passes the test since it does not involve a definitive prediction." Despite their legitimacy as falsification tests of hypotheses, falsifiable retrodictions are qualitatively inferior to falsifiable predictions, in Oldershaw's view:

> But, in the final analysis, only true definitive predictions can justify the promotion of a theory from being viewed as one of many plausible hypotheses to being recognized as the best available approximation of how nature actually works. A theory that cannot generate definitive predictions, or whose definitive predictions are impossible to test, can be regarded as inherently untestable.

A less sympathetic view concerning the validity of retrodiction as a scientific tool was offered by Gee,[44] who dismissed the legitimacy of all historical hypotheses on the ground that "they can never be tested by experiment, and so they are unscientific.... No science can ever be historical." This viewpoint, in turn, has been challenged by Cleland,[45] who contends that "when it comes to testing hypotheses, historical science is not inferior to classical experimental science" but simply exploits the available evidence in a different way:

> There [are] fundamental differences in the methodology used by historical and experimental scientists. Experimental scientists focus on a single (sometimes complex) hypothesis, and the main research activity consists in repeatedly bringing about the test conditions specified by the hypothesis, and controlling for extraneous factors that might produce false positives and false negatives. Historical scientists, in contrast, usually concentrate on formulating multiple competing hypotheses about particular past events. Their main research efforts are directed at searching for a smoking gun, a trace that sets apart one hypothesis as providing a better causal explanation (for the observed traces) than do the others. These differences

> in methodology do not, however, support the claim that historical science is methodologically inferior, because they reflect an objective difference in the evidential relations at the disposal of historical and experimental researchers for evaluating their hypotheses.

Cleland's approach has the merit of preserving as "scientific" some of the most important hypotheses advanced in such historical fields of inquiry as geology, evolutionary biology, cosmology, paleontology, and archaeology. As Cleland has noted:

> Experimental research is commonly held up as the paradigm of successful (a.k.a.good) science. The role classically attributed to experiment is that of testing hypotheses in controlled laboratory settings. Not all scientific hypotheses can be tested in this manner, however. Historical hypotheses about the remote past provide good examples. Although fields such as paleontology and archaeology provide the familiar examples, historical hypotheses are also common in geology, biology, planetary science, astronomy, and astrophysics. The focus of historical research is on explaining existing natural phenomena in terms of long past causes. Two salient examples are the asteroid-impact hypothesis for the extinction of the dinosaurs, which explains the fossil record of the dinosaurs in terms of the impact of a large asteroid, and the "big-bang" theory of the origin of the universe, which explains the puzzling isotropic three-degree background radiation in terms of a primordial explosion. Such work is significantly different from making a prediction and then artificially creating a phenomenon in a laboratory.[46]

In a paper presented to the 2004 Astrobiology Science Conference (Cleland, 2004), Cleland extended this analytic framework to the consideration of putative biosignatures as evidence of the past or present existence of extraterrestrial life. Acknowledging that "because biosignatures represent indirect traces (effects) of life, much of the research will be historical (vs. experimental) in character even in cases where the traces represent recent effects of putative extant organisms," Cleland concluded that it was appropriate to employ the methodology that characterizes successful historical research:

> Successful historical research is characterized by (1) the proliferation of alternative competing hypotheses in the face of puzzling evidence and (2) the search for more evidence (a "smoking gun") to discriminate among them.[47]

From the perspective of the evidentiary standards applicable to historical science in general and astrobiology in particular, the key retrodiction of the SB hypothesis—that the fundamental constants of nature that comprise the Standard Model as well as other physical features of our cosmos (including the number of extended physical dimensions and the extremely low value of dark energy) will be collectively bio-friendly—appears to constitute a legitimate scientific test of the hypothesis. Moreover, within the framework of Goal 7 of the NASA Astrobiology Roadmap, the retrodicted biophilic quality of our universe appears, under the SB hypothesis, to constitute a possible biosignature.

10. Caution Regarding the Use of Retrodiction to Test the SB Hypothesis

Because the SB hypothesis is radically novel and because the use of falsifiable retrodiction as a tool to test such an hypothesis creates at least the appearance of a "confirmatory argument[t] resemble[ing] just-so stories (Rudyard Kipling's fanciful stories, e.g., how leopards got their spots)" [48] it is important (as noted previously) that two additional criteria be satisfied before this retrodiction can be considered a legitimate test of the hypothesis:

- The SB hypothesis must generate falsifiable predictions as well as falsifiable retrodictions; and
- The SB hypothesis must be consilient with key theoretical constructs in such "adjoining" areas of scientific investigation as M-theory, cosmogenesis, complexity theory, and evolutionary theory.

As argued at length elsewhere,[49] the SB hypothesis is both consilient with central concepts in these "adjoining" fields and fully capable of generating falsifiable predictions.

11. Concluding Remarks

In his book *The Fifth Miracle*[50] Davies offered this interpretation of NASA's view that the presence of liquid water on an alien world was a reliable marker of a life-friendly environment:

> In claiming that water means life, NASA scientists are... making—tacitly—a huge and profound assumption about the nature of nature. They are saying, in effect, that the laws of the universe are cunningly contrived to coax life into being against the raw odds; that the mathematical principles of physics, in their elegant simplicity, somehow know in advance about life and its vast complexity. If life follows from [primordial] soup with causal dependability, the laws of nature encode a hidden subtext, a cosmic imperative, which tells them: "Make life!"

> And, through life, its by-products: mind, knowledge, understanding. It means that the laws of the universe have engineered their own comprehension. This is a breathtaking vision of nature, magnificent and uplifting in its majestic sweep. I hope it is correct. It would be wonderful if it were correct. But if it is, it represents a shift in the scientific world-view as profound as that initiated by Copernicus and Darwin put together.

An emerging consensus among mainstream physicists and cosmologists is that the particular universe we inhabit appears to confirm what Smolin calls the "anthropic observation": the laws and constants of nature seem to be fine-tuned, with extraordinary precision and against enormous odds, to favor the emergence of life and its byproduct, intelligence. As Dyson put it eloquently more than two decades ago:

> The more I examine the universe and study the details of its architecture, the more evidence I find that the universe in some sense must have known that we were coming. There are some striking examples in the laws of nuclear physics of numerical accidents that seem to conspire to make the universe habitable.[51]

Why this should be so remains a profound mystery. Indeed, the mystery has deepened considerably with the recent discovery of the inexplicably tiny value of dark energy density and the realization that M-theory encompasses an unfathomably vast landscape of possible solutions, only a minute fraction of which correspond to anything resembling the universe that we inhabit.

Confronted with such a deep mystery, the scientific community ought to be willing to entertain plausible explanatory hypotheses that may appear to be unconventional or even radical. However, such hypotheses, to be taken seriously, must:

- be consilient with the key paradigms of "adjoining" scientific fields,
- generate falsifiable predictions, and
- generate falsifiable retrodictions.

The SB hypothesis satisfies these criteria. In particular, it generates a falsifiable retrodiction that the physical laws and constants that prevail in our cosmos will be biophilic—which they are.

Appendix B

COEVOLUTION OF THE COSMIC PAST AND FUTURE: THE SELFISH BIOCOSM AS A CLOSED TIMELIKE CURVE

by James N. Gardner

As published in the May/June 2005 issue
of *Complexity* magazine, vol. 10, no. 5
reprinted with permission

Abstract

The Selfish Biocosm (SB) hypothesis asserts that the anthropic qualities which our universe exhibits can be explained as incidental consequences of a cosmic replication cycle in which a cosmologically extended biosphere supplies two of the essential elements of self-replication identified by von Neumann. It was previously suggested (1) that the hypothesis implies that the emergence of life and intelligence are key epigenetic thresholds in the cosmic replication cycle, strongly favored by the physical laws and constants which prevail in our particular universe and (2) that a falsifiable implication of the hypothesis is that the emergence of increasingly intelligent life is a robust phenomenon, strongly favored by the natural processes of evolution which result from the interplay of those laws and constants. Here I propose a cosmic evolutionary paradigm, consistent with both the SB hypothesis and the modified ekpyrotic cyclic universe scenario, by means of which a life-friendly suite of physical laws and constants could conceivably emerge. The key feature of the proposed paradigm is the constrained coevolution of past and future cosmic states along a hypothesized closed timelike curve.

Keywords: Anthropic, biocosm, closed timelike curve, cosmology, ekpyrotic.

1. Introduction

In a paper published in this journal,[1] I first advanced the hypothesis that the anthropic qualities which our universe exhibits can be explained as incidental consequences of a cosmic replication cycle in which the emergence of a cosmologically extended biosphere could conceivably supply two of the logically essential elements of self-replication identified by John von Neumann[2]: a controller and a duplicating device. The hypothesis advanced in "The Selfish Biocom" was an attempt to extend and refine Lee Smolin's conjecture[3] that the majority of the anthropic qualities of the universe can be explained as incidental consequences of a process of cosmological replication and natural selection (CNS) whose utility function is black hole maximization. Smolin's conjecture differs crucially from the concept of eternal chaotic inflation advanced by Andrei Linde[4] in that it predicts a cosmological evolutionary process with a specific and discernible utility function—black hole maximization. It is this aspect of Smolin's conjecture rather than the specific utility function he advocates that renders his theoretical approach genuinely novel.

As noted previously,[5] Smolin's conjecture suffers from two evident defects: (1) the fundamental physical laws and constants do not, in fact, appear to be fine-tuned to favor black hole maximization and (2) no mechanism is proposed corresponding to two logically required elements of any von Neumann self-replicating automaton: a controller and a duplicator.

Theories of cosmological eschatology previously articulated[6] predict that the ongoing process of biological and technological evolution is sufficiently robust and unbounded that, in the far distant future, a cosmologically extended biosphere could conceivably exert a global influence on the physical state of the cosmos. A related set of insights from complexity theory indicates that the process of emergence resulting from such evolution is essentially unbounded.

A synthesis of these two sets of insights yields the two key elements of the Selfish Biocosm (SB) hypothesis. The essence of that synthesis is that the ongoing process of biological and technological evolution and emergence could conceivably function as a von Neumann controller and that a cosmologically extended biosphere could, in the very distant future, function as a von Neumann duplicator in a hypothesized process of cosmological replication.

In "Assessing the Robustness of the Emergence of Intelligence" it was suggested that a falsifiable implication of the SB hypothesis is that the process of the progression of the cosmos through critical epigenetic thresholds in its life cycle, while perhaps not strictly inevitable, is relatively robust. One such critical threshold is the emergence of human-level and higher intelligence, which is essential to the eventual scaling up of biological and technological processes to the stage at which those processes could conceivably exert a global influence on the state of the cosmos. Four specific tests of the robustness of the emergence of human-level and higher intelligence were proposed.[7]

In "Assessing the Computational Potential of the Eschaton" it was proposed that an additional falsifiable implication of the SB hypothesis is that there exists a plausible final state of the cosmos that exhibits maximal computational potential. This predicted final state appeared to be consistent with both the modified ekpyrotic cyclic universe scenario[8] and with Seth Lloyd's description of the physical attributes of the ultimate computational device: a computer as powerful as the laws of physics will allow.[9]

2. Other Perspectives on a Hypothetical Cosmic Replication Cycle Mediated by Life and Intelligence

The central assertions of the SB hypothesis are: (1) that highly evolved life and intelligence play a central role in the process of cosmic replication and (2) that the peculiarly life-friendly laws and physical constants that prevail in our universe—an extraordinarily improbable ensemble that the late Heinz Pagels dubbed the cosmic code—play a cosmological role functionally equivalent to that of DNA in an earthly organism: they provide a recipe for cosmic ontogeny and a blueprint for cosmic reproduction.

Related topics have been seriously considered by only a handful of scientists. The late British astronomer Fred Hoyle, a self-proclaimed atheist famous for the now discredited theory of a steady-state universe but less well known for his astonishing predictions concerning the anthropic aspects of stellar nucleosynthesis inside giant supernovae, concluded that the most straightforward explanation for the astonishing array of life-friendly coincidences embedded in the laws and constants of nature was that a superintellect (but not a supernatural intellect) located somewhere in space and time had somehow deliberately engineered the laws of physics to make it possible for carbon-based life and intelligence to evolve. In a 1982 essay[10] he explained this possibility:

> Would you not say to yourself, "Some super-calculating intellect must have designed the properties of the carbon atom, otherwise the chance of my finding such an atom through the blind forces of nature would be utterly minuscule?" Of course you would. A common sense interpretation of the facts suggests that a superintellect has monkeyed with physics, as well as with chemistry and biology, and that there are no blind forces worth speaking about in nature. The numbers one calculates from the facts seem to me so overwhelming as to put this conclusion almost beyond question.

The unknown superintelligence that preceded us, Hoyle believed, put together as a "deliberate act of creation" a universe that was suitable for carbon-based life and the evolution of intelligence. Hoyle stressed that the superintellect of which he was speaking was not a supernatural deity but a natural entity whose essence we could ultimately aspire to understand. Indeed, far from being religiously inspired, Hoyle's idea that such a naturally occurring superintellect might have existed and might have been responsible for the deliberate engineering of the basic laws of nature was, in his view, deeply antithetical to the proreligion bias of Western civilization and culture.

Another contemporary scientist who has articulated similar ideas is the astronomer Edward Harrison. In an audacious scientific paper published in Britain in 1995 in the *Quarterly Journal of the Royal Astronomical Society*,[11] Harrison suggested that our universe was created by life-forms possessing superior intelligence existing in another physical universe in which the constants of physics were finely tuned and therefore essentially similar to our own. Further, Harrison suggested that highly intelligent beings, perhaps including our own descendants in the far future, might possess not only the knowledge to design but also the technology to build baby universes. Finally, Harrison conjectured that the very comprehensibility of the universe to the human mind might be a subtle clue that the universe was, in fact, designed by minds basically similar to our own.

Finally, in an unpublished paper entitled "On the Role of Life in the Evolution of the Universe,"[12] Kansas State University mathematician Louis Crane suggested that Harrison's speculations required us to consider the possibility that the historical processes of biological and cosmological evolution are inseparably linked:

> In the first place, the origin and evolution of life [can] no longer [be viewed as] a mere accident. Rather it is deliberately coded into the fine tuning of the physical laws. Since the development of life and of the universe are joined into a unified evolutionary process, they can be viewed from the point of view of purpose, just as it makes sense to speak of the purpose of an organ of a developing animal, even though the development of the animal is entirely within the scope of physical law.
>
> Secondly, intelligence and its ongoing success are no longer a small and unimportant accident in an enormous universe. Rather they are the precondition for the existence and reproduction of the universe. The world around us was created by something like us, and is structured, as if deliberately, to produce us and nurture us. We have a larger purpose which goes beyond ourselves of sustaining and recreating the universe.

3. The First Cause Problem

The courageous speculations of Harrison, Hoyle, Crane, and a handful of other iconoclastic scientists unfortunately shed no light on a central problem raised by the SB hypothesis: Who or what initially launched the process of cosmic reproduction? How did life and intelligence first become possible? How did the "first" universe become sentient and thus capable of seeding its progeny?

The tests of the SB hypothesis previously proposed do not address the issue of how the information governing the process of biological evolution and emergence that is hypothesized to lead up to the cosmological replication event can plausibly arise in the first instance. Yet in order for the hypothesis to be credible, there must exist a plausible natural mechanism by means of which the specified complexity that is the essence of biological information can initially arise. Put differently, since the hypothesis implies that the origin of life and intelligence is a cosmic imperative, encoded as a subtext to the laws and constants of inanimate nature, it follows as a falsifiable implication of that hypothesis there must be a plausible natural means through which that life-friendly cosmic code was initially composed. But what could that natural process possibly be? Who or what could have first authored the cosmic code?

Faced with the daunting—and directly analogous—puzzle of life's origin, Charles Darwin famously opined[13] that this particular challenge, no less than the "extreme difficulty or rather impossibility of conceiving this immense and wonderful universe, including man with his capacity of looking far backwards and far into futurity, as the result of blind chance or necessity," was perhaps forever beyond the reach of what he called humanity's godlike intellect. Equally intimidated by the origin-of-life puzzle, DNA codiscoverer Francis Crick sought intellectual solace in the speculative concept of "directed panspermia"—the notion that life might be able to deliberately perpetuate and multiply itself throughout the cosmos by means of a kind of intentional diaspora.[14]

The problem is that this proposed mechanism, while increasing the odds that life could permeate the cosmos, merely postpones the problem of the origin of biological information. Crick's hypothesis offers an exotic yet plausible response to the dilemma posed by physicist Enrico Fermi about the troublesome lack of evidence of extraterrestrial life (Fermi asked, If the origin and evolution of life throughout the universe is chemically predestined, why is it we have acquired no direct evidence of the hypothesized phenomenon?) by asserting that all earthly organisms may, in fact, be the remote progeny of extraterrestrial microbes. Yet the hypothesis does not seek to probe the central mystery of the origin of the primal coding mechanism and thus of the origin of life considered as the origin of specified complexity. In short, directed panspermia is a laudable attempt to explain life's dispersion but not its commencement. Viewed as a proffered solution to the origin puzzle, Crick's hypothesis must be disqualified as what physicist John Wheeler would call an inadmissible "tower of turtles standing one on the other"[15]—a framework of ideas that indulges in infinite regress and thus sidesteps the ultimate issue.

The problem of infinite regress is precisely the same whether the issue is the origin of organic life or the origin of an entire intelligently designed universe governed by a life-friendly cosmic code. Harrison characterized the dilemma as follows:

> The creation [of such] a universe . . . requires a high level of intelligence, thus raising the question of how the first universe began. A parallel problem concerns the origin of life on Earth. The probability of life originating on a planet is exceedingly small. No doubt the Galaxy teems with lifeless planets on which the conditions were never favourable for the origin of life. Life originated on Earth because of its unusual conditions.... In cosmogenesis, one possibility—a variation on the anthropic principle—is that an initial ensemble of universes, in which the fundamental parameters have random variations, contains at least one member in which intelligent life is possible. This member is the "intelligent" mother universe. Thereafter, by reproduction, intelligent universes dominate the ensemble, and the original unintelligent members then form a vanishingly small fraction of the whole. As on Earth, life originates with the first self-reproducing molecule, produced by chance, and thereafter it proliferates and dominates. . . . We are still left with the ultimate question:...who created the initial cosmic ensemble in the anthropic principle?... Does the prospect of infinite regress mean

> the discussion has now moved into realms of reality where creative agents exist in universes no longer comprehensible to the human mind?[16]

4. The Ekpyrotic Cyclic Universe as a Closed Timelike Curve

As noted earlier, [17] I suggested that the recently proposed modified ekpyrotic cyclic universe scenario offered a plausible physical template for the process of cosmological replication that is the key event under the SB hypothesis. In particular, I suggested that a falsifiable implication of the hypothesis—that there exists a plausible final state of the cosmos that exhibits maximal computational potential—appeared to be consistent with the ekpyrotic cyclic scenario's prediction of the characteristics of the final state of a cyclic cosmos immediately prior to and during a Big Bounce era and also with Lloyd's description of the ultimate computational device: a computer as powerful as the laws of physics will allow.

For purposes of the present inquiry into possible ultimate sources of the cosmic code, it is useful to summarize another predicted characteristic of the ekpyrotic cyclic scenario: the uninterrupted flow of time across the threshold of the Big Bounce era. In this scenario "the Universe undergoes an endless sequence of cosmic epochs which begin with the Universe expanding from a 'big bang' and end with the Universe contracting to a 'big crunch.'"[18] The most important aspect of that scenario for purposes of the present inquiry is that it predicts that time (and thus causation and the flow of information) continues smoothly across the Big Crunch/Big Bang era. This prediction opens up a radically novel potential venue for the origin of biological information: a coevolutionary matrix of past and future cosmic states.

In 1997, Princeton astrophysicists J. Richard Gott III and Li-Xin Li posed this intriguing question: Could our universe conceivably have spawned itself?[19] Beginning with the recognition that "a remarkable property of [Einstein's theory of] general relativity is that it allows solutions that have closed timelike curves (CTCs)"—hypothetical configurations of space and time where gravity is sufficiently strong to bend the space-time continuum into a looping configuration that allows future events to influence the past—Gott and Li pointed out that, absent some rule like the chronology protection conjecture proposed by Stephen Hawking (which states that the laws of physics conspire to forbid the actual manifestation of CTCs, at least at the macroscopic scale), the "Universe can be its own mother." Under the CTC cosmological scenario "the Universe neither tunneled from nothing, nor arose from a singularity; it created itself."

Commenting on the potential relationship of this CTC scenario to the conjecture by Edward Harrison that our life-friendly universe might be the artifact of a prior advanced civilization—a baby universe created "in the lab" by some supercivilization in a prior universe—Gott and Li noted that Harrison was able to explain multiple generations of artificially created baby universes by this mechanism, except for the first one. As Gott and Li put it, "This seems to be an unfortunate gap. In our scenario, suppose that first universe simply turned out to be one of the infinite ones formed later by intelligent civilizations. Then the Universe—note capital U—would be multiply connected, and would have a region of CTCs; all of the individual universes would owe their birth to some intelligent civilization in particular in this picture."

The ekpyrotic cyclic universe scenario adds to the conjecture put forward by Gott and Li—that the Universe might be its own mother—the crucially important possibility that time (and thus causation and information) might flow smoothly from a Big Crunch era (an epoch characterized, as previously shown, by maximal computational potential) to a Big Bang era and that the two eras might be linked by means of a CTC. This implies that the point of ultimate origin of biological information—of specified complexity—might plausibly lie in a coevolutionary matrix of past and future temporal states, causally linked by the unbounded loop of a CTC.

5. Causation from the Super-Copernican Perspective

The principal reason that Darwin doubted the capacity of the human mind to probe profound cosmic mysteries like these was his skepticism that "the mind of man, which has, as I fully believe, been developed from a mind as low as that possessed by the lowest animal, [can] be trusted when it draws such grand conclusions."[20] Always the cautious skeptic, Darwin never forgot that, to paraphrase the concluding passage of *The Descent of Man*, mankind's godlike intellect as well as his bodily frame still bore the indelible imprint of humanity's lowly origin.

However, Darwin's own monumental achievements and those of countless other pioneers of science provide ample reason to believe that such limitations are not insurmountable obstacles to human comprehension of even the most counterintuitive physical phenomena. The utterly counterintuitive but highly successful theories of quantum physics and relativity are particularly noteworthy in this regard.

For purposes of the present inquiry, the key perspective is offered by what physicist John Wheeler calls the super-Copernican principle. Derived from the Copenhagen interpretation of quantum physics, this "principle rejects the now-centeredness of any account of existence as firmly as Copernicus rejected here-centeredness."[21] According to this principle, the future can have at least as important a role in shaping the present moment as the past. The most important aspect of Wheeler's insight is not that we must embrace the specific mechanism of retroactive causation favored by Wheeler and the advocates of the Copenhagen interpretation of quantum mechanics (the retroactive impact on quantum phenomena of observer-participancy), but rather that we should be open to counterintuitive notions of causation, if they appear to be consistent with novel yet mathematically plausible accounts of physical reality.

In particular, the vision of the cosmos as a closed timelike curve that allows at least limited information flow across the putative Big Bounce threshold offers a new paradigm that may allow us to formulate radically novel theoretical possibilities concerning the origin and nature of biological information and of the specified complexity it exhibits. According to this paradigm, the process of biological information generation can be viewed as an essentially eternal autocatalytic process in which past and future temporal states are linked in a coevolutionary relationship. The wave of causation moves from what we call the past to what we call the future and back again to the past across the Big Crunch era to a new Big Bang era without disruption (but, we shall see shortly, with possible causal filtering),

Causation defines the relationship between all points on the CTC, but the relationship of cause and effect is not temporally restricted in the sense we naively perceive. As Wheeler put it with uncanny prescience (though with a different causal mechanism in mind), the history of the cosmos "is not a history as we usually conceive history. It is not one thing happening after another after another. It is a totality in which what happens 'now' gives reality to what happened 'then,' perhaps even determines what happened then."[22] Because the CTC is curved *and* timelike *and* closed *and* unblemished by a final singularity, each point on the CTC is, to at least a limited degree, both the cause *and* effect of every other point. Time flows in only one direction in this scenario but because the CTC unites past and future at the Big Crunch threshold, the two temporal states can coevolve.

The CTC that is hypothesized to be our cosmos thus may be a classic autocatalytic set, what Wheeler ventured to call a "self-excited circuit" and a "grand synthesis, pulling itself together all the time as a whole."[23] The implication for the origin of biological information should be apparent: not only the universe but also the life-friendly cosmic code and indeed life itself (and the specified complexity it embodies) could conceivably be its own mother under this scenario.

Speculating on this possibility (but without the benefit of the specific scenario provided by Gott, Li, and the modified ekpyrotic cyclic universe proposal), astrophysicist Paul Davies had this to say in an online *Edge* interview:[24]

> QUESTION: You mention aliens. Who are the aliens?
>
> DAVIES: We don't know. We could be totally alone in the universe; at this particular time it's impossible to say. But we can speculate that there might be life, even intelligent life, elsewhere.
>
> QUESTION: Could they be our ancestors? Or our God?
>
> ANSWER: Descendants maybe, not ancestors. Well, I guess if it's possible to travel through time as well as through space, we can imagine the universe being populated by a single species far into the future and also backwards into the past, so they could also be our ancestors too. It wouldn't be necessary to have life popping up independently in many different places. That would be a curious twist on the time-travel story. We could go backwards in time and seed other planets with life at an earlier epoch. Yes, that's always conceivable.

6. A Limited Chronology Protection Conjecture

Under the scenario outlined above, only *information*—not a physical entity engages in what Davies would call time travel. Only *information* makes the journey from past to future and back to past across the Big Bounce threshold. And, if a key speculation by Andrei Linde is correct, the *only* information capable of making this extraordinary journey is the cosmic code itself.

In an audacious scientific paper entitled "The Hard Art of Universe Creation" published in 1991,[25] Linde asked whether it might be theoretically possible for the fabricators of a new baby universe created "in the lab" to send a message to future living creatures who might one day inhabit such an artificially created cosmos. The only possibility for accomplishing this feat, Linde concluded, would be to embed the message in the laws of physics that would prevail in the new universe. He theorized:

> It seems that the only way to send a message to whose who will live in the universe we are planning to create is to encrypt it into the properties of the vacuum state of the new universe, i.e., into the laws of the low-energy physics. Hopefully, one may achieve it by choosing a proper combination of temperature, pressure and external fields, which would lead to creation of the universe in a desirable phase state.

Thus, a limited form of Stephen Hawking's chronology protection conjecture is preserved by Linde's speculation: no time travelers are permitted. Only a very limited form of time-traveling information is allowed.

7. Consilience with M-Theory-Inspired Cosmology

The consilience of Linde's scenario with M-theory-inspired cosmology is striking. In a brilliant insight whose import has not been widely acknowledged, Linde realized that what is normally viewed as a *weakness* of M-theory-inspired cosmology—that it does not uniquely predict a single set of laws and constants of low-energy physics but is rather capable of generating an enormous and astonishingly variegated landscape of seemingly arbitrary sets of such physical rules, the vast majority of which would not even remotely resemble the laws and constants that prevail in our universe—is, in fact, a crucial *strength* if one of the possible functions of those physical laws and constants is to transmit information to a baby universe. As Linde put it:

> The corresponding message can be long and informative enough only if there are extremely many ways of symmetry breaking and/or patterns of compactification in the underlying theory. This is exactly the case, e.g., in the superstring theory, which was considered for a long time as one of the main problems of this theory. Another requirement to the informative message is that it should not be too simple. If, for example, masses of all particles would be equal to each other, all coupling constants would be given by 1, etc., the corresponding message would be too short. Perhaps, one may say quite a lot by creating a universe in a strange vacuum state.... The stronger is the symmetry breaking, the more "unnatural" are relations between parameters of the theory after it, the more information the message may contain. Is it the reason why we must work so hard to understand

> strange features of our beautiful and imperfect world? Does this mean that our universe was created not by a divine designer but by a physicist hacker? If it is true, then our results indicate that he did a very difficult job. Hopefully, he did not make too many mistakes.

The analogy to the key property of DNA that makes it capable of encoding a construction plan for an organism is quite precise. It is the very fact that the sequence of four nucleotides (adenosine, cytosine, guanine, and thymine) in a strand of DNA is not prefigured by the inherent chemical properties of the DNA molecule that allows DNA to function as a superb genetic coding mechanism. If the letters in the DNA alphabet could not be sequenced in an arbitrary order corresponding to the "recipe" for a particular organism, then DNA would be a woefully inadequate vehicle for encoding and transmitting genetic information.

So too, as Linde has demonstrated, it is the inherent flexibility of M-theory-inspired cosmology—a conceptual framework theoretically capable of generating a whole menagerie of wildly different universes exhibiting an arbitrary mix of disparate physical laws and constants, only a tiny subset of which would be life-friendly—that makes it theoretically possible for those laws and constants actually manifested in a particular universe to encode a kind of "message," which is the functional counterpart to the "message" encoded in an earthly organism's DNA.

8. Comparison with the WAP/Eternal Chaotic Inflation Model

It is useful to briefly compare the cosmological model described in this paper with the principal alternative on offer from M-theory-inspired cosmologists.

This approach, born of intellectual desperation on the part of Leonard Susskind and other M-theory advocates, is to overlay eternal chaotic inflation with an explanatory approach known as the weak anthropic principle. The weak anthropic principle merely states in tautological fashion that since human observers inhabit this particular universe, it must perforce be life-friendly or it would not contain any observers resembling ourselves. Eternal chaotic inflation, invented by Linde, asserts that instead of just one Big Bang there are, always have been, and always will be, countless numbers of Big Bangs detonating constantly in inaccessible cosmic domains. These Big Bangs create countless numbers of new universes constantly. The whole ensemble of universes constitutes a multiverse.

The motivation for this exercise was the discovery that M-theory allows a vast landscape of possible vibration modes of superstrings, only a tiny fraction of which correspond to anything like the sub-atomic particle world we observe and that is described by the Standard Model of particle physics.

Just how big is this landscape of possible alternative models of particle physics allowed by M-theory? According to Susskind, the mathematical landscape is truly gigantic, with different and distinct environments "measured not in the millions or billions but in googles or googleplexes," none of which appears to be mathematically favored, let alone foreordained by the theory.[26] And in virtually none of those other mathematically permissible environments would matter and energy have possessed the qualities that are necessary for stars, galaxies, and carbon-based living creatures to have emerged from the primordial chaos.

This is, as Susskind notes, an intellectual cataclysm of the first magnitude because it seems to deprive our most promising new theory of fundamental physics—M-theory—of the power to uniquely predict the emergence of anything remotely resembling our universe. As Susskind puts it, the picture of the universe that is emerging from the deep mathematical recesses of M-theory is not an "elegant universe" at all. Rather it is a Rube Goldberg device, cobbled together by some unknown process in a supremely improbable manner that just happens to render the whole ensemble miraculously fit for life. In the words of University of California theoretical physicist Steve Giddings, "No longer can we follow the dream of discovering the unique equations that predict everything we see, and writing them on a single page. Predicting the constants of nature becomes a messy environmental problem. It has the complications of biology."

In an attempt to cope with the dilemma posed by the failure of M-theory to uniquely predict anything resembling our cosmos, Susskind and other M-theorists propose that in each hypothesized Big Bang that occurs in the process of eternal chaotic inflation, the laws, constants and the physical dimensionality of nature come out differently. In some, dark energy is stronger. In some, dark energy is weaker. In some, gravity is stronger. In some, gravity is weaker. In some, there are three extended spatial dimensions while in others there are as many as seven. This variation occurs, according to M-theory-based cosmology, because the 10-dimensional physical shapes in which superstrings vibrate—known as Calabi-Yau shapes—evolve randomly and chaotically at the moment of each new Big Bang. The laws and constants of nature are constantly reshuffled by this process, like a cosmic deck of cards.

On extraordinarily rare occasions this random process of eternal chaotic inflation is thought to yield a new baby universe where the prevailing physical laws and dimensionless constants are life-friendly. That outcome will be pure chance—one lucky roll of the dice in an unimaginably vast cosmic crap shoot with a googleplex of unfavorable outcomes for every winning turn.

Our universe is viewed by these theorists as a big winner in the cosmic lottery. Here is how the eminent Nobel laureate Steve Weinberg explained this scenario in a *New York Review of Books* essay[27] a couple of years ago: "The expanding cloud of billions of galaxies that we call the big bang may be just one fragment of a much larger universe in which big bangs go off all the time, each one with different values for the fundamental constants." It is no more a mystery that our particular branch of the multiverse exhibits life-friendly characteristics, according to Weinberg, than that life evolved on the hospitable Earth "rather than some horrid place, like Mercury or Pluto."

There appear to be at least three principal problems with the Weinberg/Susskind approach of overlaying the weak anthropic principle on the eternal chaotic inflation model.

First, universes spawned by Big Bangs other than our own are inaccessible from our own universe, at least with the experimental techniques currently available to science. So the approach appears to be untestable, perhaps untestable in principle. And testability is the hallmark of genuine science, distinguishing it from fields of inquiry like metaphysics and theology.

Second, the Weinberg/Susskind approach extravagantly violates the mediocrity principle. The mediocrity principle, a mainstay of scientific theorizing since Copernicus, is a statistically based rule of thumb that, absent contrary evidence, a particular sample (Earth, for instance, or our particular universe) should be assumed to be a typical example of the ensemble of which it is a part. The Weinberg/Susskind approach flagrantly flouts the mediocrity principle. Instead, their approach simply takes refuge in a brute, unfathomable mystery—the conjectured lucky roll of the dice in a crap game of eternal chaotic inflation—and declines to probe seriously into the possibility of a naturalistic cosmic evolutionary process that has the capacity to yield a life-friendly set of physical laws and constants on a nonrandom basis.

Third, the Weinberg/Susskind approach needlessly inflates the probabilistic resources required to explain the phenomenon of a life-friendly cosmic code. The alternative explanation offered in this paper is considerably more parsimonious because it assumes minimalistically the existence of only one universe rather than the multitude of universes required under the WAP/eternal chaotic inflation model. (To be precise, the SB hypothesis permits but does not require the existence of a multiverse consisting of an ensemble of eternally reproducing baby and mother universes.)

9. The Constrained Coevolution of Past and Future Cosmic States

The cosmic picture that emerges from this paper is a paradigm of constrained past state/future state coevolution, reminiscent of Stuart Kauffman's notion[28] that life originated from the interaction of autocatalytic sets of carbon-based polymers. The new picture requires us to reconceive the phenomenon of ultimate cosmic causation from a fresh perspective in which neither the past nor the future is causally privileged or primary. Under this paradigm, past and future states of the cosmos coevolve, at least to a limited degree, with the putative Big Bounce serving as a kind of semi-porous causation filter. Only the cosmic code is transmitted across the Big Bounce threshold from cosmic cycle to cosmic cycle, much as the information encoded in DNA (but no additional biological information) is transmitted from one generation of earthly creatures to the next. Subject to the constraints of this filter, past and future cosmic states thus comprise a classic autocatalytic set.

This is a radically novel picture of cosmic causation and, indeed, a startlingly counterintuitive vision of the universe—an image of a surpassingly strange cosmos, presciently described by John Wheeler as a "self-excited circuit" and a "grand synthesis, pulling itself together all the time as a whole." It is a cosmological paradigm characterized preeminently by the appearance of "order for free"—an informational matrix that autocatalyzes its own emergence by linking past and future states of the cosmos in a coevolutionary relationship.

While strange and counterintuitive, this paradigm is scarcely more bizarre than the notion of a universe that fluctuates into existence *ex nihilo* at the moment of the Big Bang (and thus constitutes, in the phrase of cosmologist Alan Guth, the ultimate "free lunch") or a universe in which the supposedly absolute speed limit imposed by the theory of relativity on the propagation of causal effects (the speed of light in a vacuum) is seemingly violated with impunity by the experimentally confirmed phenomenon of quantum non-locality. These phenomena remind us that we inhabit a universe that is, in the inimitable phrase of John B. S. Haldane, not only queerer than we imagine but queerer than we can possibly imagine.

Appendix C

The Selfish Biocosm: Complexity as Cosmology

by James N. Gardner

As published in the January/February 2000 issue
of *Complexity* magazine, vol. 5, no. 3
reprinted with permission

In Vital Dust Nobel laureate Christian de Duve issued a daunting challenge to biologists and philosophers seeking to unify their seemingly incommensurable intellectual realms:

> Traditionally, the dialogue with philosophers has been held mainly by theoretical physicists and mathematicians, probably because of a common meeting ground in abstraction. The resulting cosmological picture comprised all facets of the physical world, from elementary particles to galaxies, but either ignored life or had life and mind tagged on to the picture as separate entities by some implicit, sometimes explicit, recourse to vitalism and dualism. *This is wrong. Life is an integral part of the universe; it is even the most complex and significant part of the know universe. The manifestations of life should dominate our world picture, not be excluded from it. This has become particularly mandatory in view of the revolutionary advances in our understanding of life's fundamental processes.*[1]

In the five years that have elapsed since the publication of de Duve's call for cross-disciplinary dialogue, the first hints of a revolutionary new biology-centered cosmological model have begun to emerge. This new paradigm, foreshadowed in earlier speculations by Stephen Hawking, John Wheeler, Freeman Dyson, Fred Hoyle and other eminences, rests on a sober assessment of the astonishing array of "just so" coincidences inherent in the physical characteristics of our universe–characteristics which render the cosmos peculiarly friendly to carbon-based life.

One key example is described by Hawking in an essay entitled "Quantum Cosmology":

> The trouble with the hot big bang model is the trouble with all cosmology that has no theory of initial conditions: it has no predictive power. Because general relativity would break down at a singularity, anything could come out of the big bang. So why is the universe so homogeneous and isotropic on a large scale, yet has local irregularities such as galaxies and stars? Any why is the universe so close to the dividing line between collapsing again and expanding indefinitely? *In order to be as close as we are now, the rate of expansion early on had to be chosen fantastically accurately. If the rate of expansion one second after the big bang had been less by one part in 10^{10}, the universe would have collapsed after a few million years. If it had been greater by one part in 10^{10}, the universe would have been essentially empty after a few million years. In neither case would it have lasted long enough for life to develop. Thus one either has to appeal to the anthropic principle or find some physical explanation of why the universe is the way it is.*[2]

Many other examples of cosmological oddities which render the universe eerily hospitable to organic life are analyzed in The Anthropic Cosmological Principle by astronomer John Barrow and physicist Frank Tipler.[3] The central point, according to Barrow and Tipler, is that:

> ...it is not only man that is adapted to the universe. The universe is adapted to man. Imagine a universe in which one or another of the fundamental dimensionless constants of physics is altered by a few percent one way or the other? Man could never come into being in such a universe. That is the central point of the anthropic principle. According to the principle, a life-giving factor lies at the center of the whole machinery and design of the world.[4]

What is one to make of this spooky set of coincidences? Reactions vary widely among distinguished scientists. Some, like the late Heinz Pagels, disparage anthropic explanations as impediments to the quest for a final cosmological theory that would embed the apparently arbitrary constants of nature in an elegant and self-evident set of truly fundamental physical rules.[5] Others, like the iconoclastic physicist John Wheeler, deride efforts to derive the laws of nature and the values of physical constants from time-invariant principles:

> Surely–big bang and gravitational collapse advise us–the laws of physics cannot have existed from everlasting to everlasting. They must have come into being at the one gate in time, must fade away at the other. But at the beginning there were no gears and pinions, no corps of Swiss watchmakers to put things together, not even a preexisting plan. If this assessment is correct, every law of physics must be at bottom like the second law of thermodynamics, higgledy-piggledy in character, based on blind chance.[6]

As the 21st Century looms, the approach of mavericks like Wheeler appears to be gaining momentum.

Freeman Dyson offered this prediction in 1985:

> I am suggesting that there may come a time when physics will be willing to learn from biology as biology has been willing to learn from physics, a time when physics will accept the endless diversity of nature as one of its central themes, just as biology has accepted the unity of the genetic coding apparatus as one of its central dogmas.[7]

The time foreseen by Dyson, I believe, is rapidly approaching. With the advent of the new biologically focused scientific paradigm of self-organizing complexity and with the publication of groundbreaking cosmological speculations like physicist Lee Smolin's The Life of the Cosmos,[8] a serious effort to undertake a grand unification of biology and cosmology appears to be commencing. Even skeptics like physicist Steven Weinberg–who concluded his popular account of the "first three minutes" with the extravagantly gloomy observation that "[t]he more the universe seems comprehensible, the more it also seems pointless"[9]–have been swept along by the new intellectual tide. As reported in a recent issue of Scientific American:

> One direction, explored recently by Steven Weinberg of the University of Texas at Austin and his colleagues, invokes the last resort of cosmologists, the anthropic principle. If the observed universe is merely one of an infinity of disconnected universes–each of which might have slightly different constants of nature, as suggested by some incarnations of inflationary theory combined with emerging ideas of quantum gravity–then physicists can hope to estimate the magnitude of the cosmological constant by asking in which universes intelligent life is likely to evolve.[10]

This essay will offer, in admittedly coarse-grained fashion, a variation on the theme of Lee Smolin's biology-derived cosmological paradigm. My goal is to commence a serious exploration of the possibility foreseen in the 1950s by chemist Michael Polanyi that "the universe is still dead, but it already has the capacity of coming to life."[11]

The Cosmos-as-Replicator Concept

Selective cosmological replication is the heart of Lee Smolin's concededly speculative hypothesis. As Smolin himself has acknowledged, what is genuinely novel about his theory is not the specific utility function purportedly maximized by the hypothesized process of selective cosmic replication but rather the notion of the universe as a self-organizing and self-reproducing replicator, competing for proliferation success within a set of cosmic replicators possessing disparate powers of replication:

> I believe more in the general idea that there must be mechanisms of self-organization involved in the selection of the parameters of the laws of nature than I do in this particular mechanism, which is only the first one I was able to invent.[12]

Thus, while cosmologists have seriously questioned whether the particular utility function hypothesized by Smolin–black hole production–is indeed maximized by the laws and constants of nature which prevail in our particular universe,[13] they have not thereby undermined the essence of Smolin's new paradigm: his vision of the universe as a self-organizing replicator which competes for reproductive success within a multiverse of cosmic replicators possessing varying degrees of replicator power.

However, there is a more fundamental flaw in Smolin's hypothesis, which may be characterized as the problem of memory. For purposes of this analysis, I use the term "memory" in the broad sense defined by Gerald Edelman in Bright Air, Brilliant Fire:

> I submit that * * * memory * * * takes many forms but has general characteristics that are found in all its variations. I am using the word "memory" here in a more inclusive fashion than usual. Memory is a process that emerged only when life and evolution occurred and gave rise to the system described by the sciences of recognition. As I am using the term memory, it describes aspects of heredity, immune responses, reflex learning, true learning following perceptual categorization, and the various forms of consciousness. In these instances, structures evolved that permit significant correlations between current ongoing dynamic patterns and those imposed by past patterns. These structures all differ, and memory takes on its properties

> as a function of the system in which it appears. What all memory systems have in common is evolution and selection. Memory is an essential property of biologically adaptive systems.[14]

Formulation of an hypothesis suggesting a means by which the *memory* (in Edelman's sense) of a particular cosmos could conceivably arise and persist through the process of cosmic replication is a crucial link missing from Smolin's theory. As John Baez has observed:

> Smolin's theory is based on two hypotheses.
>
> A. The formation of a black hole creates "baby universes," the final singularity of the black hole tunneling right on through to the initial "big bang" singularity of the new universe thanks to quantum effects. While this must undoubtedly seem *outré* to anyone unfamiliar with the sort of thing theoretical physicists amuse themselves with these days, in a recent review article by John Preskill on the information loss paradox for black holes, he reluctantly concluded that this was the "most conservative" solution of that famous problem.
>
> * * *
>
> B. Certain parameters of the baby universe are close to but different than those of the parent universe. The notion that certain physical facts that appear as "laws" are actually part of the state of the universe has in fact been rather respectable since the application of spontaneous symmetry breaking to the Weinberg-Salam model of electroweak interactions, part of the standard model.* * * So again, while the idea must seem wild to anyone who has not encountered it before, physicists these days are fairly comfortable with the idea that certain "fundamental constants" could have been other than they were. *As for the constants of a baby universe being close to, but different than, those of the parent universe, there is as far as I know no suggested mechanism for this. This is perhaps the weakest link in Smolin's argument.*[15]

Richard Dawkins made the same point in a recent exchange with Smolin:

> Note that any Darwinian theory depends on the prior existence of the strong phenomenon of heredity. There have to be self-replicating entities (in a population of such entities) that spawn daughter entities more like themselves than the general population.[16]

Dawkins' observation echoes, at least in part, the conclusions of John von Neumann in his 1948 Caltech lecture entitled "On the General and Logical Theory of Automata," which have been summarized as follows:

What von Neumann discovered was that *any* self-reproducing object must contain four fundamental components:

A. A *blueprint*, providing the plan for construction of offspring

B. A *factory*, to carry out the construction

C. A *controller*, to ensure that the factory follows the plan

D. A *duplicating machine*, to transmit a copy of the blueprint to the offspring.[17]

In the context of Smolin's hypothesis, one can surmise that the physical laws and constants of our universe and its presumed progeny could conceivably constitute a von Neumann blueprint (literally a "cosmic code," in Pagels' phrase[18]) and the universe at large could serve as a sort of von Neumann factory.[19] But what device or process could play the roles of von Neumann controller or von Neumann duplicating machine? It may turn out to be the case, as Martin Rees has written, that "[t]he mechanisms that might 'imprint' the basic laws and constants in a new universe are obviously far beyond anything that we can understand."[20] But unless Rees is mistaken and unless Baez's critique can be answered plausibly, Smolin's hypothesis appears to be untenable *ab initio*.

Emergence as an Element of Cosmological Theory

Could the sciences of complexity come to the rescue of Smolin's theory and suggest, at least in principle, a set of processes that could supply the two missing von Neumann elements of a self-reproducing system? Possibly.

Smolin himself has predicted that the paradigms underlying the sciences of complexity will come to play an ever-larger role in cosmology:

> [I]t seems to me quite likely that the concept of self-organization and complexity will more and more play a role in astronomy and cosmology. I suspect that as astronomers become more familiar with these ideas, and as those who study complexity take time to think seriously about such cosmological puzzles as galaxy structure and formation, a new kind of astrophysical theory will develop, in which the universe will be seen as a network of self-organized systems. Beyond this, I also think that * * * this merging of the science of the fundamental and the science of the organized will overturn the usual ways of thinking about the elementary particles, too. Many of the people who work on complexity * * * imagine that the world consists of highly organized and complex systems but that the fundamental laws are simply fixed beforehand, by God or by mathematics. I used to believe this, but I no longer

> do. *More and more, what I believe must be true is that there are mechanisms of self-organization extending from the largest scales to the smallest, and that they explain both the properties of the elementary particles and the history and structure of the whole universe.*[21]

One key phenomenon associated with self-organization and the concept of complex adaptive systems is emergence, described succinctly by John Holland:

> The hallmark of emergence is this sense of much coming from little.* * * We are everywhere confronted with emergence in complex adaptive systems–ant colonies, networks of neurons, the immune system, the Internet, and the global economy, to name a few–where the behavior of the whole is much more complex than the behavior of the parts.[22]

Four generic features of the phenomenon of emergence are noteworthy in the context of the present inquiry. First, as Holland notes, "[t]he possibilities for emergence are compounded when the elements of the system include some capacity, however elementary, for adaptation or learning."[23] Second, the "component mechanisms [in an emergent system] interact without central control."[24] Third, the "possibilities for emergence increase rapidly as the flexibility of the interactions increases."[25] Fourth, and perhaps most important, "persistent patterns at one level of observation can become building blocks at still more complex levels,"[26] yielding a defining characteristic of emergent systems as embodying "*hierarchical organization* (configurations of generators become generators at a higher level of organization)."[27]

The fourth characteristic of emergent systems is especially crucial because it implies that the number of hierarchical levels underlying a particular emergent phenomenon can be indefinitely large and that sufficiently complex multilevel hierarchies of basic components as simple as quarks and sub-atomic particles (the initial products of the Big Bang) are, in proper combination, capable of eventually yielding such high-level phenomena as human culture (including specialized domains of that culture like scientific inquiry):

> [H]uman creative activity, ranging from the construction of metaphors through innovations in business and government to the creation of new scientific theories, seems to involve a controlled invocation of emergence. We are everywhere confronted with emergence in complex adaptive systems–ant colonies, networks of neurons, the immune system, the Internet, and the global economy, to name a few–where the behavior of the whole is much more complex than the behavior of the parts.[28]

The same point about the indefinitely large hierarchical layering potential of emergent complex adaptive systems was made implicitly by Murray Gell-Mann in the inaugural issue of this journal:

> Examples on Earth of the operation of complex adaptive systems include biological evolution, learning and thinking in animals (including people), the functioning of the immune system in mammals and other vertebrates, the operation of the human scientific enterprise, and the behavior of computers that are built or programmed to evolve strategies–for example by means of neural nets or genetic algorithms. *Clearly, complex adaptive systems have a tendency to give rise to other complex adaptive systems.*[29]

There is no indication that the "tendency" noted by Gell-Mann has been arrested at this particular historical moment. On the contrary, the evidence is overwhelming that the process of multi-level hierarchical emergence is accelerating rapidly.

How far can the process of emergence propel the phenomenon of complexification[30] in theory? How high can this process allow mankind and its progeny to "climb Mt. Improbable" (in Richard Dawkins' felicitous phrase[31]), employing only the ropes and pitons furnished by the principles of complexity theory and evolution? Several recent speculations about the potential magnitude of the ongoing process of cosmological emergence are worth noting as a prelude to formulating a tentative answer to these portentous questions.

Cosmological Emergence Scenario # 1: The Kurzweil Vision

Ray Kurzweil's The Age of Spiritual Machines[32] offers a plausible prophecy succinctly conveyed by the subtitle of his book: "When Computers Exceed Human Intelligence." When exactly will that be? While other artificial intelligence theorists like Gerald Edelman speculate that we may someday be capable of constructing what Edelman calls a "conscious artifact,"[33] most computer scientists believe that such an extraordinary technological feat lies far in the future.

Not Kurzweil. The computer pioneer forecasts that by 2020, advanced computers will exceed the memory capacity and computational ability of the human brain, with human-like attributes of synthetic emotion and natural speech not far behind. A mere ten years later, Kurzweil predicts, human brains will be linked seamlessly with their electronic counterparts, allowing information to flow directly between ourselves and our artificial progeny. Not long thereafter, machines will gain decisive advantage over their creators as the inexorable logic of quickening technological innovation (which Kurzweil encapsulates in a general principle he calls the Law of Time and Chaos) drives their intellectual capacities far beyond ours. As Kurzweil puts it, "Once a computer achieves a human level of intelligence, it will necessarily roar past it."[34]

The debate over whether machines can be endowed with consciousness is scarcely novel. Over the years scientists like Alan Turing and Roger Penrose as well as philosophers like Daniel C. Dennett and John Searle have debated the issue ad nauseam. What is original about Kurzweil's contribution is that he places the anticipated emergence of superior machine intelligence squarely in the context of biological evolution. As Kurzweil puts it:

> Evolution has been seen as a billion-year drama that led inexorably to its grandest creation: human intelligence. The emergence in the early twenty-first century of a new form of intelligence on Earth that can compete with, and ultimately significantly exceed, human intelligence will be a development of greater import than any of the events that have shaped human history. It will be no less important than the creation of the intelligence that created it, and will have profound implications for all aspects of human endeavor, including the nature of work, human learning, government, warfare, the arts, and our concept of ourselves.[35]

And what are the ultimate prospects for such machines? In Kurzweil's view, they may prove capable of cosmological engineering on the grandest scale:

> [H]ow relevant is intelligence to the rest of the Universe? The common wisdom is, *Not very.* ⋆ ⋆ ⋆ The Universe itself was born in a big bang and will end with a crunch or a whimper; we're not yet sure which. But intelligence has little to do with it. Intelligence is just a bit of froth, an ebullition of little creatures darting in and out of inexorable universal forces. The mindless mechanism of the Universe is winding up or down to a distant future, and there's nothing intelligence can do about it. That's the common wisdom. But I don't agree with it. My conjecture is that intelligence will ultimately prove more powerful than these big impersonal forces. ⋆ ⋆ ⋆ The implication of the Law of Accelerating Returns is that intelligence on Earth and in our Solar System will vastly expand over time. The same can be said across the galaxy and throughout the Universe. It is likely that our planet is not the only place where intelligence has been seeded and is growing. Ultimately, intelligence will be a force to reckon with, even for these big celestial forces (so watch out!). The laws of physics are not repealed by intelligence, but they effectively evaporate in its presence. So will the Universe end in a big crunch, or in an infinite expansion of dead stars, or in some other manner? In my view, the primary issue is not the mass of the Universe, or the possible existence of antigravity, or of Einstein's so-called cosmological constant. Rather, the fate of the Universe is a decision yet to be made, one which we will intelligently consider when the time is right.[36]

Cosmological Emergence Scenario # 2: The Wheeler Vision

John Wheeler's vision of the distant future has been dubbed the *participatory anthropic principle*. It was summarized by Wheeler in a presentation to the Santa Fe Institute in 1989, later published in an essay entitled "It From Bit." Wheeler had speculated in an earlier lecture that while "[t]he anthropic principle superficially looks like a tautology,"[37] it may in fact be subject to genuine prediction and thus "to destruction in the sense of Karl Popper."[38] With this assumption in mind, Wheeler proceeded to consider a possible explanation of the now-famous "just so" coincidences which make carbon-based life and intelligence possible in our universe:

> Is the machinery of the universe so set up, and from the very beginning, that it is guaranteed to produce intelligent life at some long-distant point in its history-to-be? And is this proposition testable * * * ? Perhaps. But how should such a fantastic correlation come about between big and small, between and [sic] machinery and life, between future and past? * * * [H]ow can history ever have made things come out right, ever given a world of life, ever thrown up a communicating community of the kind required for the establishment of meaning? In brief, how can the machinery of the universe ever be imagined to get set up at the very beginning so as to produce man now? Impossible! Or impossible unless somehow–preposterous idea–meaning itself powers creation. But how? Is that what the quantum is all about?[39]

In "It From Bit," Wheeler expanded on this speculation:

> *It from bit.* Otherwise put, every it–every particle, every field of force, even the space-time continuum itself–derives its function, its meaning, its very existence entirely–even if in some contexts indirectly–from the apparatus-elicited answers to yes or no questions, binary choices, *bits.* It from bit symbolizes the idea that every item of the physical world has at bottom–at a very deep bottom, in most instances–an immaterial source and explanation; that what we call reality arises in the last analysis from the posing of yes-no questions, and the registering of equipment-evoked responses; in short, that all things physical are information-theoretic in origin and this is a *participatory universe.*[40]

Three elements of Wheeler's vision of a participatory universe are particularly relevant for purposes of the current analysis: (1) his concept of the cosmos as an autocatalytic loop ("To endlessness [infinite regress] no alternative is evident but loop, such a loop as this: Physics gives rise to observer-participancy; observer-participancy given rise to information; and information gives rise to physics."[41]); (2) his notion of the universe, not as a machine, but as self-organized system ("Directly opposed to the concept of universe as machine built on law is the vision of a world self-synthesized. On this view, the notes struck out on a piano by the observer-participants of all places and all times, bits though they are, in and by themselves constitute the great wide world of space and time and things."[42]); and (3) what Wheeler calls the "super-Copernican" principle:

> The super-Copernican principle. This principle reject now-centeredness in any account as firmly as Copernicus rejected here-centeredness. It repudiates most of all any tacit adoption of here-centeredness in assessing observer-participants and their numbers. ⋆ ⋆ ⋆ We today, to be sure, through our registering devices, give a tangible meaning to the history of the photon that started on its way from a distant quasar long before there was any observer-participancy anywhere. However the far more numerous establishers of meaning of time to come have a like inescapable part–by device-elicited questions and registration of answer–in generating the "reality" of today. For this purpose, moreover, there are billions of years yet to come, billions on billions of sites of observer-participancy yet to be occupied. How far foot and ferry have carried meaning-making communication in fifty thousand years gives faint feel for how far interstellar propagation is destined to carry it in fifty billion years.[43]

A final "clue" noted by Wheeler ties his theory tightly to one of the key concepts underlying the sciences of complexity:

> *Fifth and final clue*: More is different. ⋆ ⋆ ⋆ We do not have to turn to objects so material as electrons, atoms, and molecules to see big numbers generating new features. The evolution from small to large has already in a few decades forced on the computer a structure reminiscent of biology by reason of its segregation of different activities into distinct organs. Distinct organs, too, the giant telecommunications system of today finds itself inescapably evolving. Will we someday understand time and space and all the other features that distinguish physics–and existence itself–as the similarly self-generated order of a self-synthesized information system?[44]

Cosmological Emergence Scenario # 3: The Barrow/Tipler Vision

In the concluding chapter of The Anthropic Cosmological Principle,[45] John Barrow and Frank Tipler set forth what they call the Omega Point[46] theory of the final state of the universe. They begin their analysis by noting that while "there is no evidence whatsoever of intelligent life having any significant effect upon the Universe in the large,"[47] this may not always be the case:

> We *know* space travel is possible. We argued in Chapter 9 that even interstellar travel is possible. Thus once space travel begins, there are, in principle, no further physical barriers to prevent *Homo sapiens* (or our descendants) from eventually expanding to colonize a substantial portion, if not all, of the visible Cosmos. Once this has occurred, it becomes quite reasonable to speculate that the operations of all these intelligent beings could begin to affect the large scale evolution of the Universe. If this is true, it would be in this era–in the far future near the Final State of the Universe–that the true significance of life and intelligence would manifest itself. Present-day life would then have cosmic significance because of what future life may someday accomplish.[48]

Barrow and Tipler proceed to speculate that in the far-distant future, the boundary of the biosphere will expand to be coterminous with that of the cosmos itself:

> Finally, a time is reached when life has encompassed the entire Universe and regulated all matter contained therein. Life begins to manipulate the dynamical evolution of the universe as a whole, forcing the horizons to disappear, first in one direction, and then another. The information stored continues to increase * * *. From our [prior] discussion * * * we see that if life evolves in all of the many universes in a quantum cosmology, and if life continues to exist in all of these universes, then all of these universes, which include all possible histories among them, will approach the Omega Point. At the instant the Omega Point is reached, life will have gained control of all matter and forces not only in a single universe, but in all universes whose existence is logically possible; life will have spread into all spatial regions in all universes which could logically exist, and will have stored an infinite amount of information, including all bits of knowledge which it is logically possible to know. And this is the end.[49]

The authors conceive of one possible "reaso[n] to think that life is essential to the Cosmos"[50] but do not address the possibility that the life processes might eventually be capable of the ultimate feat of cosmological engineering: cosmic replication.

Cosmological Emergence Scenario # 4: The Dyson Vision

In Infinite in All Directions Freeman Dyson offers a vision of the distant future similar in many respects to the Barrow/Tipler scenario. Like the preceding speculators, Dyson places life and intelligence at the center of any serious inquiry into the ultimate fate of the cosmos:

> It is impossible to calculate in detail the long-range future of the universe without including the effects of life and intelligence. It is impossible to calculate the capabilities of life and intelligence without touching, at least peripherally, philosophical questions. If we are to examine how intelligent life may be able to guide the physical development of the universe for its own purposes, we cannot altogether avoid considering what the values and purposes of intelligent life may be.[51]

Dyson contrasts his vision with the nihilistic observation of Weinberg quoted previously, foreseeing a universe ever more suffused with life, intelligence and purpose:

> The universe that I have explored in a very preliminary way ⋆ ⋆ ⋆ is very different from the universe which Weinberg envisaged when he called it pointless. I have found a universe growing without limit in richness and complexity, a universe of life surviving forever and making itself known to its neighbors across unimaginable gulfs of space and time. Whether the details of my calculations turn out to be correct or not, there are good scientific reasons for taking seriously the possibility that life and intelligence can succeed in molding this universe of ours to their own purposes.[52]

Dyson even divines a candidate "law of nature" from the tendency of conscious thought to exert ever greater control over inanimate matter:

> To me the most astounding fact in the universe ⋆ ⋆ ⋆ is the power of mind which drives my fingers as I write these words. Somehow, by natural processes still totally mysterious, a million butterfly brains working together in a human skull have the power to dream, to calculate, to see and to hear, to speak and to

> listen, to translate thoughts and feelings into marks on paper which other brains can interpret. Mind, through the long course of biological evolution, has established itself as a moving force in our little corner of the universe. Here on this small planet, mind has infiltrated matter and has taken control. It appears to me that the tendency of mind to infiltrate and control matter is a law of nature.[53]

The operation of this "law of nature," Dyson believes, implies that life and intelligence will play a dominant role in shaping the physical eschatology of the cosmos:

> Individual minds die and individual planets may be destroyed. But, as Thomas Wright said, "The catastrophy of a world, such as ours, or even the total dissolution of a system of worlds, may possibly be no more to the great Author of Nature, than the most common accident of life with us." The infiltration of mind into the universe will not be permanently halted by any catastrophe or any barrier that I can imagine. If our species does not choose to lead the way, others will do so, or may have already done so. If our species is extinguished, others will be wiser or luckier. Mind is patient. Mind has waited for 3 billion years on this planet before composing its first string quartet. It may have to wait for another 3 billion years before it spreads all over the galaxy. I do not expect that it will have to wait so long. But if necessary, it will wait. The universe is like a fertile soil spread out all around us, ready for the seeds of mind to sprout and grow. Ultimately, late or soon, mind will come into its heritage.[54]

What use will life and intelligence make of this "fertile soil"? Dyson is deeply skeptical about the capacity of our inherently bounded human intellects to probe this ultimate mystery:

> What will mind choose to do when it informs and controls the universe? This is a question which we cannot hope to answer. When mind has expanded its physical reach and biological organization by many powers of ten beyond the human scale, we can no more expect to understand its thoughts and dreams than a Monarch butterfly can understand ours. ★ ★ ★ In contemplating the future of mind in the universe, we have exhausted the resources of our puny human science. This is the point at which science ends and theology begins.[55]

Cosmological Emergence Scenario # 5: The Dawkins Vision

The work of evolutionary theorist Richard Dawkins is not ordinarily associated with cosmological models or physical eschatology. Nonetheless, his notion of ascending hierarchies of replicators pursuing emergent categories of replication objectives furnishes a valuable conceptual tool with which to synthesize the insights of the four theoreticians discussed above.

Dawkins' vision of the hierarchical layering of emergent replicator categories is stated most clearly in the concluding chapter of River Out of Eden:

> There is another type of explosion [besides a supernova explosion] a star can sustain. Instead of "going supernova" it "goes information." The explosion begins more slowly than a supernova and takes incomparably longer to build up. We can call it an information bomb or, for reasons that will become apparent, a replication bomb. For the first few billion years of its build-up, you could detect a replication bomb only if you were in the immediate vicinity. Eventually, subtle manifestations of the explosion begin to leak away into more distant regions of space and it becomes, at least potentially, detectable from a long way away. We do not know how this kind of explosion ends. Presumably it eventually fades away like a supernova, but we do not know how far it typically builds up first. Perhaps to a violent and self-destructive catastrophe. Perhaps to a more gentle and repeated emission of objects, moving, in a guided rather than a simple ballistic trajectory, away from the star into distant reaches of space, where it may infect other star systems with the same tendency to explode.[56]

Human life, in Dawkins view, plays an important catalytic role in the "detonation" of the replication bomb:

> We humans are an extremely important manifestation of the replication bomb, because it is through us–through our brains, our symbolic culture and our technology–that the explosion may proceed to the next stage and reverberate through deep space.[57]

The commencement of the process, however, antedates the arrival of the human race by billions of years:

> We have no direct evidence of the replication event that initiated the proceedings on this planet. We can only infer that it must have happened because of the gathering explosion of

> which we are a part. We do not know exactly what the original critical event, the initiation of self-replication, looked like, but we can infer what kind of an event it must have been. It began as a chemical event. ★ ★ ★ What, then, was this momentous critical event that began the life explosion? I have said that it was the arising of self-duplicating entities, *but equivalently we could call it the origination of heredity–a process of "like begets like."*[58]

Dawkins proposes a replicator classification system which comprises a detailed hierarchy of "replicator thresholds," beginning with Threshold 1 (the "Replicator Threshold" itself)[59] continuing through Threshold 5 (the "High-Speed Information Processing Threshold"), Threshold 6 (the "Consciousness Threshold"), Threshold 7 (the "Language Threshold") and Threshold 8 (the "Cooperative Technology Threshold").

At some point between Threshold 5 and Threshold 8, Dawkins theorizes, an entirely new class of replicators arose: selfish, self-replicating memes:

> [I]t is possible that human culture has fostered a genuinely new replication bomb, with a new kind of self-replicating entity–the meme, as I have called it in The Selfish Gene–proliferating and Darwinizing in a river of culture. There may be a meme bomb now taking off, in parallel to the gene bomb that earlier set up the brain/culture conditions that made the take-off possible.[60]

The final replication threshold foreseen by Dawkins is Threshold 10 (the "Space Travel Threshold") which he describes as follows:

> After radio waves, the only further step we have imagined in the outward progress of our own explosion is physical space travel itself: Threshold 10, the Space Travel Threshold. Science-fiction writers have dreamed of the interstellar proliferation of daughter colonies of humans, or their robotic creations. These daughter colonies could be seen as seedlings, or infections, of new pockets of self-replicating information–pockets that may subsequently themselves expand explosively outward again, in satellite replication bombs, broadcasting both genes and memes. If this vision is ever realized, it is perhaps not too irreverent to imagine some future Christopher Marlowe reverting to the imagery of the digital river: "See, see, where life's flood streams in the firmament!"[61]

The momentous question posed by Smolin's hypothesis can be restated in terms of Dawkins' classification scheme: Is Threshold 10 truly the final replication threshold? Or might there be a Threshold 11, which we may provisionally call the Cosmic Replication Threshold? Might Threshold 11 harbor a radically new type of replicator–differing from the preceding classes as profoundly as the meme differs from the gene but incorporating the complex interactions of those precedent entities as subroutines–which we might provisionally label (in deference to Dawkins' memorable nomenclature) as the Selfish Biocosm replicator class?

Synthesis of the Five Cosmological Emergence Scenarios

We are now prepared to formulate a crude synthesis of key elements of the five cosmological emergence scenarios discussed above. This synthesis yields two dramatic hypotheses.

First, according to the first four scenarios, life is capable of attaining the capacity to engage in cosmological engineering in the very distant future. While the mechanism postulated by Wheeler (retroactive quantum mechanical effects of observer-participancy) differs from the vaguely stated technological assumptions of the other theorists, all agree on the potential magnitude of the future effect of life on the global state of the cosmos.

Second, Dawkins' open-ended replicator hierarchy suggests that the natural processes of self-organization, emergence and natural selection, governed by laws whose existence is hypothesized (but not yet definitively formulated) by complexity theorists as well as by theoretical approaches derived from Darwinian theory, are fully capable of yielding such a capability without any requirement of supernatural intervention or supervision.

The Concept of the Selfish Biocosm

We began this inquiry in an effort to determine whether the sciences of complexity could rescue Lee Smolin's daring hypothesis of selective cosmological replication by supplying two essential von Neumann components conspicuously absent from his theory: a controller and a duplicating machine.

Synthesis of the five cosmological emergence scenarios summarized above together with an admittedly speculative application of key concepts underlying the sciences of complexity yields a preliminary answer:

1. Tentatively identified principles guiding the evolution and operation of complex adaptive systems could conceivably function as a von Neumann controller, governing the process of cosmological self-organization and emergence at all relevant scales, leading up to a cosmological replication event.

2. Life itself, when it reaches a requisite threshold of pervasiveness and evolved sophistication at or near the Barrow/Tipler Omega Point, could conceivably serve as the requisite Von Neumann cosmological duplicating machine.[62]

A Disquieting Perspective

Freeman Dyson famously expounded upon the seemingly miraculous concatenation of cosmic laws and constants which render the cosmos mysteriously life-friendly, confessing that "[t]he more I examine the universe and study the details of its architecture, the more evidence I find that the universe in some sense must have known we were coming."[63] Some scientists like physicist Paul Davies hail this strong anthropic perspective as "magnificent and uplifting":

> In claiming that water means life, NASA scientists are * * * making–tacitly–a huge and profound assumption about the nature of nature. They are saying, in effect, that the laws of the universe are cunningly contrived to coax life into being against the raw odds; that the mathematical principles of physics, in their elegant simplicity, somehow know in advance about life and its vast complexity. If life follows from [primordial] soup with causal dependability, the laws of nature encode a hidden subtext, a cosmic imperative, which tells them: "Make life!" And, through life, its by-products: mind, knowledge, understanding. It means that the laws of the universe have engineered their own comprehension. This is a breathtaking vision of nature, magnificent and uplifting in its majestic sweep. I hope it is correct. It would be wonderful if it were correct. But if it is, it represents a shift in the scientific world-view as profound as that initiated by Copernicus and Darwin put together.[64]

Others will have the opposite reaction if they consider the matter carefully. From the Selfish Biocosm perspective, earthly life and human intelligence are not the grand climax of creation but rather minuscule operants in a surpassingly complex process that our particular universe employs in order to get itself grown to maturity and then reproduced. The disquieting jolt induced by this change of perspective is reminiscent of that furnished by the "meme's eye view" of human culture, characterized as follows by Daniel C. Dennett:

> This [memetic perspective] is a new way of thinking about ideas. It is also, I hope to show, a good way, but at the outset the perspective it provides is distinctly unsettling, even appalling. We can sum it up with a slogan: A scholar is just a library's way of making another library. I don't know about you, but I'm not initially attracted by the idea of my brain as a sort of dung heap in which the larvae of other people's ideas renew themselves, before sending out copies of themselves in an

> informational Diaspora. It does seem to rob my mind of its importance as both author and critic. Who's in charge, according to this vision–we or our memes?[65]

To view the multi-billion-year pageant of life's proliferation on Earth and the evolution of human intelligence as minor subroutines subordinated to inconceivably vast ontogenic processes through which our particular universe prepares itself for replication is scarcely to place mankind at the center of creation. It is rather to adopt a profoundly super-Copernican perspective (to use Wheeler's phrase). Far from offering an anthropocentric vision of the cosmos, the Selfish Biocosm perspective relegates humanity and its probable mechanical progeny to the functional equivalents of mitochondria–formerly independent biological entities whose talents were harnessed in the distant past to serve the greater good of eukaryotic ascendance.

But this emerging paradigm may offer a small measure of solace. If the Selfish Biocosm perspective is humbling, it is also infused with grandeur in precisely the sense articulated by Charles Darwin in the concluding passage of The Origin of Species:

> It is interesting to contemplate a tangled bank, clothed with many plants of many kinds, with birds singing on the bushes, with various insects flitting about, and with worms crawling through the damp earth, and to reflect that these elaborately constructed forms, so different from each other, and dependent upon each other in so complex a manner, have all been produced by laws acting around us.
>
> These laws, taken in the largest sense, being Growth with Reproduction; Inheritance which is almost implied by reproduction; Variability from the indirect and direct action of the conditions of life, and from use and disuse: a Ratio of Increase so high as to lead to a Struggle for Life, and as a consequence to Natural Selection, entailing Divergence of Character and Extinction of less-improved forms. Thus, from the war of nature, from famine and death, the most exalted object which we are capable of conceiving, namely, the production of the higher animals, directly follows. There is grandeur in this view of life, with its several powers, having been originally breathed by the Creator into a few forms or into one; and that, whilst this planet has gone cycling on according to the fixed law of gravity, from so simple a beginning endless forms most beautiful and most wonderful have been, and are being evolved.[66]

Conclusion

Superstring theorist Brian Greene speculates in The Elegant Universe that the final theory of the cosmos might take a form very different from that assumed by conventional theoreticians–a form more historically contingent, more dependent on the vagaries of evolution and emergence than upon self-evident principles of mathematical physics:

> [E]ven if Smolin's specific proposal turns out to be wrong, it does present yet another shape that the ultimate theory might take. The ultimate theory may, at first sight, appear to lack rigidity. We may find that it can describe a wealth of universes, most of which have no relevance to the one we inhabit. And moreover, we can imagine that this wealth of universes may be physically realized, leading to a multiverse–something that, at first sight, forever limits our predictive power. *In fact, however, this discussion illustrates that an ultimate explanation can yet be achieved, so long as we grasp not only the ultimate laws but also their implications for cosmological evolution on an unexpectedly grand scale.*[67]

This thought was echoed by Martin Rees:

> [W]hat we call the fundamental constants–the numbers that matter to physicists–may be *secondary consequences* of the final theory, rather than direct manifestations of its deepest and most fundamental level.[68]

In a related vein, Stephen Hawking concluded recently in a lecture entitled "Quantum Cosmology, M-Theory and the Anthropic Principle," that "the Anthropic Principle is essential, if one is to pick out a solution to represent our universe, from the whole zoo of solutions allowed by M theory."[69]

Lee Smolin's hypothesis of selective cosmological replication, supplemented by key insights drawn from complexity theory, offers a plausible and parsimonious scenario by which our universe could have developed the anthropic qualities it presently exhibits.[70] This new point of view, which I have labeled the Selfish Biocosm perspective, represents an admittedly radical paradigm shift. But perhaps such a shift is required if we are ever to realize the tantalizing possibility foreseen by John Wheeler in his 1989 Santa Fe Institute lecture:

> A single question animates this report: Can we ever expect to understand existence? Clues we have, and work to do, to make headway on that issue. Surely someday, we can believe, we will grasp the central idea of it all as so simple, so beautiful, so compelling that we will all say to each other, "Oh, how could it have been otherwise? How could we all have been so blind so long?"[71]

Notes

Dedication

1. Mitton, *Conflict in the Cosmos*, 328-329.

Foreword

1. "Smolin vs. Susskind: The Anthropic Principle."

Introduction

1. Hoyle, "The Universe: Past and Present Reflections," 8-12.
2. Crick, *Life Itself: Its Origin and Nature.*
3. Ibid., 15-16.
4. Ibid., 49.
5. Ibid., 88.
6. Dyson, *Infinite in All Directions*, 118.
7. Gardner, *Biocosm.*

Chapter 1

1. Quoted in James N. Gardner, "Ninth Circuit's Unpublished Opinions: Denial of Equal Justice?" *American Bar Association Journal* (October, 1975), 1124, 1127.
2. For a delightful account of the impact of *On the Revolutions of the Heavenly Spheres* in general and of Galileo's determination to both comply with the letter of the Vatican's censorship edict and preserve the legibility of the censored text, see Owen Gingerich, *The Book Nobody Read: Chasing the Revolutions of Nicolaus Copernicus*, 145.
3. Wolfram, *A New Kind of Science*, 465.
4. Ibid., 465–466.
5. Fredkin, "Digital Mechanics: An Informational Process Based on Reversible Universal Cellular Automata."
6. Wright, "Did the Universe Just Happen?"
7. Ibid., 17.
8. Ibid.
9. Ibid.
10. Fredkin, "On the Soul," 5.
11. Tipler, *The Physics of Immortality.*
12. Wright, "Did the Universe Just Happen?," 3.

13. Lloyd, *Programming the Universe*, 3.
14. Ibid.
15. Ibid., 3–5.
16. Schrödinger, *What Is Life?*, 20.
17. Ibid., 47.
18. Ibid., 49.
19. McFadden, *Quantum Evolution*, 101.
20. Lloyd, *Programming the Universe*, 6.
21. Ibid., 118.
22. This concept is discussed by Kauffman in two books: *At Home in the Universe* (Oxford: Oxford University Press, 1995) and *Investigations* (Oxford: Oxford University Press, 2000).
23. Lloyd, *Programming the Universe*, 169.
24. Lloyd, "A theory of quantum gravity based on quantum computation," 1.
25. Ibid., 2.
26. Wheeler, *At Home in the Universe*, 296.
27. Lloyd, "Life, the Universe, and Everything."
28. Adams, *The Hitchhiker's Guide to the Galaxy*.
29. See Gordon E. Moore's paper titled "Cramming more components onto integrated circuits," *Electronics* (April 1965), available at *download.intel.com/research/silicon/moorespaper.pdf.*
30. Kassan, "AI Gone Awry: The Futile Quest for Artificial Intelligence," 30, 33.
31. Darwin, *The Descent of Man*, 405.
32. Spector, "And now, digital evolution."
33. Unless, of course, those limits are loosened by genetic engineering. This possibility is explored at length in my first book, *Biocosm*.
34. Koza, *Routine Human-Competitive Machine Intelligence by Means of Genetic Programming.*
35. Petit, "Touched by nature: Putting evolution to work on the assembly line."
36. Gibbs, "Programming With Primordial Ooze."
37. More precisely, he wrote a series of books on the topic: *Genetic Programming I, II, III,* and *IV* (Kluwer Academic Publishers) as well as numerous monographs and lectures, many of which are available at his Website: *www.genetic-programming.com/johnkoza.html.*
38. PowerPoint slides accompanying John Koza's Accelerating Change 2003 conference talk available at *www.genetic-programming.com/256.*
39. Quoted in John R. Koza, "Routine Human-Competitive Machine Intelligence by Means of Genetic Programming," available at *www.genetic-programming.com/cofes2004.pdf.*
40. Scientific Research Interests, *www.genetic-programming.com/johnkoza.html.*
41. Ibid.
42. Penrose, *The Emperor's New Mind.*
43. Ibid., xv.
44. Ibid., xvi.
45. Penrose, *Shadows of the Mind: A Search for the Missing Science of Consciousness.*
46. Penrose, *The Emperor's New Mind*, xxii–xxiii.
47. Olson, "Interview with Seth Lloyd," November 17, 2002, quoted in Ray Kurzweil, *The Singularity Is Near*, 451. Seth Lloyd's skeptical comments are representative of the views of many quantum physicists with regard to Roger Penrose's decidedly iconoclastic hypothesis about the quantum computational role played by microtubules in the human brain in generating consciousness: "I think that it is incorrect that microtubules perform computing tasks in the brain in the way that [Penrose] and Hameroff have proposed. The brain is a hot, wet place. It is not a very favorable environment for exploiting quantum coherence. The kinds of superpositions and assembly/disassembly of microtubules for which they search do not seem to exhibit quantum entanglement.... The brain clearly isn't a classical, digital computer by any means. But my guess is that it performs most of its tasks in a 'classical' manner. If you were to take a large enough computer, and model all of the neurons, dendrites, synapses, and such, [then] you could probably get the thing to do most of the tasks that brains perform. I don't think that the brain is exploiting any quantum dynamics to perform tasks."
48. Spector, *Automatic Quantum Computer Programming.*

49. Giraldi, et al. "Genetic Algorithms and Quantum Computation."
50. Private communication to author (March 28, 2006).
51. Giraldi, et al. "Genetic Algorithms and Quantum Computation."
52. Butler, "Darwin Among the Machines" (June 13, 1863) in *The Notebooks of Samuel Butler* (Festing Jones, 1912), quoted in Kurzweil, *The Singularity Is Near*, 205.

CHAPTER 2

1. Butler, "Darwin Among the Machines" (June 13, 1863) in *The Notebooks of Samuel Butler* (Festing Jones, 1912), quoted in Kurzweil, *The Singularity Is Near*, 205.
2. Quoted in Lynn Margulis and Dorion Sagan, *What Is Life*, 1.
3. Schrödinger, *What Is Life?*, 70-71.
4. Ibid., 20.
5. James D. Watson, *DNA: The Secret of Life*, 35.
6. Ibid., xii.
7. Schrödinger, *What Is Life?*, 80.
8. Quoted in Lynn Margulis and Dorion Sagan, *What Is Life*, 2-3.
9. Bedau, "Artificial Life," 197-211.
10. Ibid.
11. For a detailed discussion of this idea of pre-biotic evolution, see Stuart Kauffman, *The Origins of Order* (Oxford: Oxford University Press, 1993).
12. Bedau, "Artificial Life."
13. *Proceedings of the Seventh International Conference on Artificial Life*, ed. Mark A. Bedau, John S. McCaskill, Norman H. Packard, and Steen Rasmussen (Cambridge: MIT Press, 2000), xi.
14. Hobbes, *Leviathan*, 1.
15. Bedau, "How to understand the question 'What is life?'," 126-129.
16. Bedau, "Four Puzzles About Life," 125, 137.
17. Bedau, "Artificial Life."
18. Ibid.
19. Crick, *Life Itself: Its Origin and Nature*, 42.
20. Ibid., 47.
21. Bedau, "How to understand the question 'What is life?,'" 126-129.
22. Hoyle, *The Black Cloud*.
23. Ibid., 156.
24. Bedau, "Four Puzzles About Life,"125, 127.
25. Bedau, "The scientific and philosophical scope of artificial life," 395-400.
26. Ibid.
27. Bedau, "Four Puzzles About Life," 128.
28. Ibid., 132.

CHAPTER 3

1. Ulum, "Tribute to John Neumann," 1-49.
2. Good, "Speculations Concerning the First Ultraintelligent Machine," 31-88.
3. Vinge, "The Coming Technological Singularity."
4. Ibid.
5. Ibid.
6. The extended terrestrial biosphere is the totality of all life on the planet plus all of life's artifacts, including the biologically perturbed atmosphere, as well as the physical and cultural output of human technological civilization. A similar but more limited concept was articulated by Richard Dawkins in *The Extended Phenotype* (Oxford: Oxford University Press, 1999 [revised edition]).
7. Tipler, *The Physics of Immortality*, 2.
8. Quoted in Ray Kurzweil, *The Singularity Is Near*, 10.
9. Kurzweil, *The Singularity Is Near*, 10-11.
10. Ibid., 7.

11. Ibid., 9.
12. Ibid.
13. Ibid.
14. Ibid., 30.
15. Vinge, "The Coming Technological Singularity."
16. Mulhall, "Are We Guardians, Or Are We Apes Designing Humans?"
17. Kurzweil, *The Singularity Is Near*, 21.
18. Reynolds, "A Rapture for the Rest of Us."
19. Kurzweil and Grossman, *Fantastic Voyage: Live Long Enough to Live Forever.*
20. Hamilton, "Chasing Immortality:—The Technology of Eternal Life: An Interview with Ray Kurzweil," 61.
21. Ibid., 65.
22. Ibid.
23. Ibid., 64.
24. Hamilton, "Even the Heavens Are Not Immortal—An Alluring Vision of Death: An Interview with Connie Barlow," 93, 94.
25. Ibid., 96.
26. Kurzweil, "Reinventing Humanity: The Future of Machine-Human Intelligence," 39, 40.
27. Ibid., 43-44.
28. Ibid.
29. Ibid., 45.
30. Ibid.
31. Kurzweil, *The Singularity Is Near*, 262-263.
32. Surowiecki, *The Wisdom of Crowds.*
33. Ibid., 17-18.
34. Ibid., 18-19.
35. Ibid., 80-81.
36. Vinge, "The Coming Technological Singularity."
37. Ibid.
38. "Themes" page of Website for *Accelerating Change 2005: Artificial Intelligence and Intelligence Anplification* (available at *www.accelerating.org/ac2005/themes.html*).
39. Ibid.
40. Raley, "Electric Thoughts?," 76, 78.
41. Dobbs, "A Revealing Reflection," 22, 24.
42. Ramachandran, "Mirror Neurons and Imitation Learning as the Driving Force Behind 'The Great Leap Forward' in Human Evolution."
43. Dobbs, "A Revealing Reflection," 22, 24.
44. Raley, "Electric Thoughts?" 76, 81.

CHAPTER 4

1. Gato-Rivera, "Brane Worlds, the Subanthropic Principle and the Undetectability Conjecture."
2. Ibid.
3. Seager, et. al. "Vegetation's Red Edge: A Possible Spectroscopic Biosignature of Extraterrestrial Plants."
4. Ibid., 2.
5. Arnold, "Transit Lightcurve Signatures of Artificial Objects."
6. Surowiecki, *The Wisdom of Crowds*, xx-xxi.
7. Ibid., xxi.
8. Ibid.
9. Tarter, "The Cosmic Haystack Is Large," 31.
10. Ibid.
11. Gato-Rivera, "Brane Worlds, the Subanthropic Principle and the Undetectability Conjecture."
12. Grinspoon, *Lonely Planets: The Natural Philosophy of Alien Life*, 315.
13. Djorgovski, "Virtual Observatory: From Concept to Implementation," 2.

14. Ibid., 5-6.
15. Gato-Rivera, "Brane Worlds, the Subanthropic Principle and the Undetectability Conjecture," 2-3.
16. Ibid., 8.
17. Ibid.
18. Dick, "Cultural evolution, the postbiological universe and SETI," 65-74.
19. Lloyd, "Ultimate Physical Limits to Computation," 1047-1054.
20. Kurzweil, *The Age of Spiritual Machines: When Computers Exceed Human Intelligence.*
21. Morris, *Life's Solution: Inevitable Humans in a Lonely Universe.*
22. See *www.seti.org/site/pp.asp?c=ktJ2J9MMIsE&b=617353.*
23. Vakoch, "Universal Translator Might be Needed to Understand ET."

CHAPTER 5

1. Gardner, "Artificial Exo-Society Modeling: A New Tool for SETI Research."
2. Epstein, "Agent-Based Computational Models and Generative Social Science," 41.
3. Davies, "Message for the curious: please phone ET, at home."
4. Tough, "How to Achieve Contact: Five Promising Strategies."
5. Chown, "Is There Anybody in There?," 30.
6. Shostak, "Does ET Use Snail Mail?"
7. Vakoch, "Altruism: A Scientific Perspective."
8. Ibid.
9. Shostak, "What Do You Say to an Extraterrestrial?"
10. Ibid.

CHAPTER 6

1. Begley, "Science Journal: Are the universe's traits random or inevitable?"
2. Penrose, *The Road to Reality*, 754.
3. Livio and Rees, "Anthropic Reasoning."
4. Ibid.
5. Ibid.
6. Morris, *Life's Solution: Inevitable Humans in a Lonely Universe*, 327.
7. Hawking, "Quantum Cosmology," 89-90.
8. Rees, *Just Six Numbers*, x.
9. Wheeler, John, foreword to *The Anthropic Cosmological Principle* by John D. Barrow and Frank J. Tipler, vii.
10. Greene, *The Elegant Universe.*
11. Giddings, comment on "The Landscape: A Talk With Leonard Susskind."
12. Weinberg, "A Designer Universe?," 47.
13. Ibid.
14. Woit, *Not Even Wrong.*
15. Ibid., 161.
16. Ibid., 256.
17. Penrose, *The Road to Reality*, 1045.
18. Darwin, *The Origin of Species*, 649.
19. Gardner, *Biocosm.*
20. Miller, *Finding Darwin's God*, 130.
21. Darwin, *The Origin of Species*, 638.
22. Rees, "Life in Our Universe and Others: A Cosmological Perspective."
23. Casti, *Paradigms Lost*, 131-132.
24. Sakai, et al. "Is it possible to create a universe out of a monopole in the laboratory?"
25. Dyson, *Disturbing the Universe*, 250.
26. Crick, *Life Itself: Its Origin and Nature*, 46-47.
27. Ibid.
28. Darwin, *The Origin of Species*, 648-649.

29. Dawkins, *The Ancestor's Tale*.
30. Ibid., 2. There is some indication in Dawkins's newest book, titled *The God Delusion* (New York: Houghton Mifflin, 2006), that he may be willing to entertain the possibility that a naturalistic process, akin in some manner to Darwinian natural selection, may be responsible for generating the life-friendly physical traits of our cosmos.

Chapter 7

1. Shermer, "Shermer's Last Law," 33.
2. Davies, "E.T. and God."
3. Ibid.
4. Ibid.
5. Davies, "Transformations in Spirituality and Religion," 51.
6. Tarter, "SETI and the Religions of the Universe."
7. Ibid., 146–147.
8. Ibid., 147.
9. Ibid., 148.
10. Mead, "God's Country? Religion and U. S. Foreign Policy," 24, 34–35.
11. Dick, *The Biological Universe*, 3.
12. Ibid.
13. Armstrong, *A History of God*.
14. Ibid., xix.
15. Ibid.
16. Ibid., 4.
17. Wilson, *Consilience*, 138.
18. Ibid.
19. Quoted in Pinker, *The Blank State*, 156.
20. Pinker, *The Blank State*, 73.
21. Ibid., 141.
22. Shermer, *How We Believe*, 162.
23. Peacocke, "The Challenge and Stimulus of the Epic of Evolution to Theology."
24. Ibid., 92.
25. Dennett, *Breaking the Spell*, 345.
26. Ibid.
27. Gardner, "Memetic Engineering."
28. Dennett, *Breaking the Spell*, 163–164.
29. Dick, "Cosmotheology."
30. Ibid., 201.
31. Crowe, *The Extraterrestrial Life Debate, 1750–1900*, xv.
32. Ibid.
33. Quoted in Steven J. Dick, *Life on Other Worlds*, 16.
34. Ibid.
35. Dick, "Cosmotheology," 201.
36. Ibid.
37. Ibid.
38. Ibid., 203.
39. Ibid.
40. Ibid., 201.
41. Ibid., 204.
42. Dennett, *Breaking the Spell*, 14.
43. Dyson, "Religion from the Outside," 6.
44. Gardner, *Biocosm*, 226.
45. Whitehead, *Science and the Modern World*, 13.
46. Ibid., 12.

47. Dick, "Cosmotheology," 204.
48. Ibid., 205.
49. Polanyi, *Personal Knowledge*, 404.

Chapter 8

1. Gardner, *Biocosm*, 9.
2. Minkel, "Hard Landscape," 21.
3. Dick, *Life on Other World*, 265.
4. Dick, "Cosmotheology, 191.
5. Monod, *Chance and Necessity*, 145-146.
6. de Duve, *Vital Dust*, 293.
7. Henderson, *The Fitness of the Environment*, 312.
8. Cirkovic, "Macro-Engineering in the Galactic Context: A New Agenda for Astrobiology."
9. Ibid., 13.
10. Ibid., 17.
11. Chaisson, *The Life Era*, 194.
12. Dyson, "Energy in the Universe," 51.
13. Dick, "Cultural evolution, the postbiological universe and SETI," 71.
14. Ibid., 69.
15. Gardner, "The physical constants as biosignature: an anthropic retrodiction of the Selfish Biocosm Hypothesis," 229-236.
16. Shostak, foreword to *Biocosm*, xxi.
17. Rees, Martin "Life in Our Universe and Others: A Cosmological Perspective."

Chapter 9

1. Gott and Li, "Can the Universe Create Itself?"
2. Mitton, *Conflict in the Cosmos*.
3. Ibid., xv-xvi.
4. Ibid., ix.
5. Ibid., xvi.
6. Ibid., xvii.
7. Ibid., xi.
8. Ibid., 141.
9. Gott, "Can the Universe Create Itself?"
10. Ibid., 1.
11. Ibid.
12. Penrose, *The Road to Reality*, 409.
13. Gott, "Can the Universe Create Itself?," 40.
14. Ibid., 41.
15. Quoted in David Brewster, *Memoirs of the Life, Writings, and Discoveries of Sir Isaac Newton: Volume 2*.
16. Penrose, *The Road to Reality*.
17. Ibid., 1033.
18. Gold, *The Deep Hot Biosphere*.
19. Carroll, "The Time Before Time."
20. Ibid.
21. Ibid.
22. Ibid.
23. Talcott, "Is Time On Our Side?," 34.
24. Yourgrau, *A World Without Time*, 134.
25. Ibid.
26. McFadden, *Quantum Evolution*, 210.
27. Ibid., 201-202.
28. Ibid., 195.

Afterword

1. Clarke, *Childhood's End.*
2. Wigner, "The Unreasonable Effectiveness of Mathematics in the Natural Sciences."
3. Ibid.
4. Ibid.
5. Ibid.
6. Schrödinger, *What Is Life?*
7. Ibid., 31.
8. Wigner, "The Unreasonable Effectiveness of Mathematics in the Natural Sciences."
9. Russell, *Study of Mathematics*, quoted in Eugene Wigner, "The Unreasonable Effectiveness of Mathematics in the Natural Sciences."
10. Gould, "Planet of the Bacteria."
11. Ibid.
12. Gould, *Full House*, 176.
13. Ibid.
14. Ibid., 178.
15. Available at *www.kurzweilai.net/meme/frame.html?m=1.*
16. Gardner, "The Selfish Biocosm: Complexity as Cosmology."
17. Ibid., 44.
18. Davies, *The Fifth Miracle*, 246.
19. Dawkins, *The Ancestor's Tale*, 613.
20. Available at *www.wie.org/j26/ultra-deep-field.asp.*
21. Dawkins, *River Out of Eden.*
22. Gardner, "The Selfish Biocosm: Complexity as Cosmology," 42.

Appendix A

1. Gardner, *Biocosm.*
2. Barrow and Tipler, *The Anthropic Cosmological Principle.*
3. Vilenkin, "Anthropic Predictions."
4. Wheeler, *At Home in the Universe.*
5. Rees, *Before the Beginning.*
6. Smolin, *The Life of the Cosmos.*
7. Greene, *The Fabric of the Cosmos.*
8. Gardner, *Biocosm.*
9. Henderson, *The Fitness of the Environment.*
10. Mitton, *Conflict in the Cosmos.*
11. Livio, "Cosmology and Life."
12. Barrow and Tipler, *The Anthropic Cosmological Principle.*
13. Goldsmith, "The Best of All Possible Worlds."
14. Bjorken, "The Classification of Universes."
15. Susskind, "The Anthropic Landscape of String Theory."
16. *www.edge.org/discourse/landscape.html.*
17. Susskind, "The Anthropic Landscape of String Theory."
18. Linde, "Inflation, Quantum Cosmology and the Anthropic Principle."
19. Weinberg, "A Designer Universe?"
20. Vilenkin, "Anthropic Predictions."
21. Hogan, "Quarks, Electrons, and Atoms in Closely Related Universes."
22. Hawking and Hertog, "Why Does Inflation Start at the Top of the Hill?"
23. Smolin, "Scientific Alternatives to the Anthropic Principle."
24. Hogan, "Quarks, Electrons, and Atoms in Closely Related Universes."
25. The metaphor furnished by the familiar process of artificial selection was Darwin's crucial stepping stone. Indeed, the practice of artificial selection through plant and animal breeding

was the primary intellectual model that guided Darwin in his quest to solve the mystery of the origin of species and to demonstrate in principle the plausibility of his theory that variation and natural selection were the prime movers responsible for the phenomenon of speciation.

26. Tipler, *The Physics of Immortality*.
27. Gardner, "The Selfish Biocosm."
28. von Neumann, "On the General and Logical Theory of Automata."
29. Smolin, *The Life of the Cosmos*.
30. Linde, "The Self-Reproducing Inflationary Universe."
31. Rees, *Before the Beginning*, 1997; Baez, online commentary.
32. Both defects were emphasized by Susskind in a recent online exchange with Smolin which appears at *www.edge.org*. Smolin has argued that his CNS hypothesis has not been falsified on the first ground but conceded that his conjecture lacks any hypothesized mechanism that would endow the putative process of proliferation of black-hole-prone universes with a heredity function: "The hypothesis that the parameters *p* change, on average, by small random amounts, should be ultimately grounded in fundamental physics. We note that this is compatible with string theory, in the sense that there are a great many string vacua, which likely populate the space of low energy parameters well. It is plausible that when a region of the universe is squeezed to Planck densities and heated to Planck temperatures, phase transitions may occur leading to a transition from one string vacua to another. But there have so far been no detailed studies of these processes which would check the hypothesis that the change in each generation is small." As Smolin noted in the same paper, it is crucial that such a mechanism exist in order to avoid the conclusion that each new universe's set of physical laws and constants would constitute a merely random sample of the vast parameter space permitted by the extraordinarily large "landscape" of M-theory-allowed solutions: "It is important to emphasize that the process of natural selection is very different from a random sprinkling of universes on the parameter space *P*. This would produce only a uniform distribution $p_{random}(p)$. To achieve a distribution peaked around the local maxima of a fitness function requires the two conditions specified. The change in each generation must be small so that the distribution can 'climb the hills' in $F(p)$ rather than jump around randomly, and so it can stay in the small volume of *P* where $F(p)$ is large, and not diffuse away. This requires many steps to reach local maxima from random starts, which implies that long chains of descendants are needed."
33. Gardner, "The Selfish Biocosm."
34. Kurzweil, *The Age of Spiritual Machines*; Dyson, *Infinite in All Directions*; Wheeler, *At Home in the Universe*.
35. Gardner, "The Selfish Biocosm."
36. Gardner, "Assessing the Robustness of the Emergence of Intelligence."
37. Gardner, "Assessing the Computational Potential of the Eschaton."
38. Khoury, Ovrut, Seiberg, Steinhardt, and Turok, "From Big Bang to Big Crunch," 2001; Steinhardt and Turok, "Cosmic Evolution in a Cyclic Universe."
39. Lloyd, "Ultimate Physical Limits to Computation."
40. Pagels, *The Cosmic Code*.
41. Wilson has identified consilience as one of the "diagnostic features of science that distinguishes it from pseudoscience" in Wilson, "Scientists, Scholars, Knaves and Fools."
42. Cleland, "Methodological and Epistemic Differences Between Historical Science and Experimental Science"; Cleland, "Historical Science, Experimental Science, and the Scientific Method"; Gee, *In Search of Deep Time*; Oldershaw, "The New Physics."
43. Oldershaw, *The New Physics*.
44. Gee, *In Search of Deep Time*.
45. Cleland, "Historical Science, Experimental Science, and the Scientific Method."
46. Cleland, "Methodological and Epistemic Differences Between Historical Science and Experimental Science."
47. Cleland, "Historical Science and the Use of Biosignatures."
48. Cleland, "Historical Science, Experimental Science, and the Scientific Method."
49. Gardner, *Biocosm*.
50. Davies, *The Fifth Miracle*.
51. Dyson, *Disturbing the Universe*.

APPENDIX B

1. Gardner, "The Selfish Biocosm," 34-45.
2. von Neumann, "On the General and Logical Theory of Automata."
3. Smolin, *The Life of the Cosmos.*
4. Linde, "The Self-Reproducing Inflationary Universe," 98-104.
5. Rees, *Before the Beginning*; Baez, J., online commentary.
6. Kurzweil, *The Age of Spiritual Machines*; Wheeler, *At Home in the Universe*; Dyson, *Infinite in All Directions.*
7. Gardner, Assessing the Robustness of the Emergence of Intelligence," 951-955.
8. Khoury, et al. "From Big Crunch to Big Bang"; Steinhardt and Turok, "Cosmic Evolution in a Cyclic Universe."
9. Lloyd, "Ultimate Physical Limits to Computation," 1047-1054.
10. Hoyle, "The Universe: Past and Present Reflections," 16-20.
11. Harrison, "The Natural Selection of Universes Containing Intelligent Life," 193-203.
12. Crane, "On the Role of Intelligent Life in the Evolution of the Universe."
13. Darwin, *Autobiography (1809-1882).*
14. Crick, *Life Itself: Its Origin and Nature.*
15. Wheeler, *Geons, Black Holes and Quantum Foam.*
16. Harrison, "The Natural Selection of Universes Containing Intelligent Life," 193-203.
17. Gardner, "Assessing the Computational Potential of the Eschaton: Testing the Selfish Biocosm Hypothesis," 285-288.
18. Steinhardt and Turok, "Cosmic Evolution in a Cyclic Universe."
19. Gott and Li, "Can the Universe Create Itself?"
20. Darwin, *Autobiography (1809-1882).*
21. Wheeler, *At Home in the Universe.*
22. Wheeler, *Geons, Black Holes and Quantum Foam.*
23. Ibid.
24. *Edge* interview with Paul Davies, *www.edge.org/3rd_culture/davies/davies_p6.html* (Dec. 8, 2002).
25. Linde, "The Hard Art of Universe Creation."
26. Susskind, "The Anthropic Landscape of String Theory."
27. Weinberg, "A Designer Universe?"
28. Kauffman, *The Origins of Order.*

APPENDIX C

1. de Duve, *Vital Dust: Life as a Cosmic Imperative*, 291.
2. Hawking, "Quantum Cosmology," in *The Nature of Space and Time*, 89-90.
3. Barrow and Tipler, *The Anthropic Cosmological Principle*, 1988.
4. Ibid, vii.
5. See Heinz R. Pagels, *Perfect Symmetry* (Bantam paperback edition, 1986), 377-379. The version of the anthropic principle disparaged by Pagels has been characterized as the *weak* anthropic principle. But what Pagels failed to note is that the anthropic principle comes in at least two flavors: the *strong* version (which asserts that the origin of life is predestined) and the *weak* version (which merely states, in tautological fashion, that since humans could only exist in a bio-friendly cosmos, the cosmos which we inhabit must perforce be bio-friendly). Ironically, Pagels seems to favor the more controversial strong version: "My own view is that although we do not yet know the fundamental laws, when and if we find them the possibility of life in a universe governed by those laws will be written into them. The existence of life in the universe is not a selective principle acting upon the laws of nature; rather it is a consequence of them." For a detailed discussion of the arguments of Pagels, see John D. Barrow, *The World Within the World* (Oxford 1988).
6. Wheeler, *At Home in the Universe*, 187.
7. Dyson, *Infinite in All Directions*, 47.
8. Smolin, *The Life of the Cosmos.*
9. Weinberg, *The First Three Minutes*, 154.

10. Krauss, "Cosmological Antigravity," 53, 59. Weinberg's endorsement of anthropic notions is decidedly qualified and appears to be restricted to a use of the weak version of the anthropic principle to explain the value of the cosmological constant. See Steven Weinberg, "A Designer Universe?" in *The New York Review of Books*, 46 et seq.
11. Polanyi, *Personal Knowledge*, 404.
12. Smolin, "A Theory of the Whole Universe," 8.
13. See Martin Rees, *Before the Beginning*, 251.
14. Edelman, *Bright Air, Brilliant Fire*, 203-204.
15. Baez, commentary on *The Life of the Cosmos.*
16. Dawkins, "Implications of Natural Selection and The Laws of Physics," 3.
17. Casti, *Paradigms Lost*, 131-132.
18. See Heinz R. Pagels, *The Cosmic Code* (Bantam paperback ed. 1983) ("What is the universe? Is it a great 3-D movie in which we are all unwilling actors? Is it a cosmic joke, a giant computer, a work of art by a Supreme Being, or simply an experiment?... I think the universe is a message written in code, a cosmic code, and the scientist's job is to decipher that code.")
19. See Heinz R. Pagels, *Perfect Symmetry*, supra note 5, at 372: "In this metaphor of the universe as a cosmic computer the material things in the universe, the quantum particles, are the 'hardware.' The logical rules these particles obey, the laws of nature, are the 'software.' The universe as it evolves can be viewed as executing a 'program' specified by the laws of nature although it is not a deterministic program like those in digital computers. What the ultimate 'output' of this cosmic computer will be remains to be determined. But we already know that its program has given rise to complex 'subroutines' that we can identify with life."
20. Rees, *Before the Beginning*, supra note 13, at 250.
21. Smolin, "A Theory of the Whole Universe," supra note 12, at 8-9.
22. Holland, *Emergence: From Chaos to Order*, 2.
23. Ibid., 5.
24. Ibid., 7.
25. Ibid.
26. Ibid.
27. Ibid., 9 (emphasis in original).
28. Ibid., 2.
29. Gell-Mann, "What Is Complexity?" 16, 17.
30. For a useful review of the concept of complexification, see John Casti, *Complexification* (HarperCollins, 1994). For an impassioned dissent from the notion that evolution embodies a trend toward increasing complexity, see Stephen Jay Gould, *Full House* (Crown, 1996). Curiously, Gould exempts *cultural* evolution from his contrarian paradigm. Id. at 217 et seq.
31. Dawkins, *Climbing Mount Improbable.*
32. Kurzweil, *The Age of Spiritual Machines.*
33. Edelman, supra note 14, at 188 et seq.
34. Kurzweil, supra note 32, at 3.
35. Kurzweil, supra note 32, at 5.
36. Ibid., 258-260.
37. Wheeler, supra note 6, at 185.
38. Ibid., 186.
39. Ibid., 185-86.
40. Ibid., 296.
41. Ibid., 300.
42. Ibid., 300-01.
43. Ibid., 305-07.
44. Ibid., 308.
45. Barrow and Tipler, supra note 3.
46. The authors define the omega point as "the future c-boundary of a universe with no event horizons [which] must consist of just a single point." Id. at 638. They acknowledge that "[i]t is possible for a closed universe to have an omega point but for life to die out before it is

reached" and distinguish between such cases and those in which it does not "by capitalizing 'Omega Point' if life reaches the omega point." Id. at 675.

47. Ibid., 613.
48. Ibid., 614 (emphasis in original).
49. Ibid., 675-77.
50. Ibid., 674-75.
51. Dyson, supra note 7, at 99-100.
52. Ibid., 117.
53. Ibid., 118.
54. Ibid.
55. Ibid., 118-19.
56. Dawkins, *River Out of Eden*, 135-36.
57. Ibid., 136-37.
58. Ibid., 138-39 (emphasis added).
59. Ibid., 151.
60. Ibid., 158. For a detailed discussion of gene/meme coevolution, see James N. Gardner, "Genes Beget Memes and Memes Beget Genes: Modeling a New Catalytic Closure, " *Complexity* (vol. 4, no. 5, May/June 1999), 22 et seq.
61. Ibid., at 160.
62. Two scientists who have put forward similar speculations are Edward Harrison and Louis Crane. See Edward Harrison, "The Natural Selection of Universes Containing Intelligent Life," Q. J. R. astr. Soc. (1995), 36, 193-203; Edward Harrison, "Creation and Fitness of the Universe," Astronomy and Geophysics (April 1998, vol. 39), p. 2.27; and Louis Crane, "Possible Implications of the Quantum Theory of Gravity: An Introduction to the Meduso-Anthropic Principle," (electronically published at *http://xxx.lanl.gov/ftp/hep-th/papers/9402/9402104.gz*).
63. Dyson, *Disturbing the Universe*, 250.
64. Davies, *The Fifth Miracle*, 246.
65. Dennett, *Consciousness Explained*, 202.
66. Darwin, *The Origin of Species*, 648-49.
67. Greene, *The Elegant Universe*, 370 (emphasis added).
68. Rees, supra note 13, at 254.
69. Hawking, "Quantum Cosmology, M-Theory and the Anthropic Principle" (*www.damtp.cam.ac.uk/user/hawking/quantum.html*). Speculation concerning the precise technology by means of which highly evolved life (the hypothetical Von Neumann copying machine) actually *copies* the "cosmic code" and *transmits* it to a daughter universe is beyond the scope of this essay. However, Hawking's remarks suggest that a fruitful area of inquiry might be M-theory and in particular the implications of M-theory with respect to the period of seemingly random evolution of Calabi-Yau shapes in the earliest post-Big Bang era. One possibility: the deliberate engineering of such shapes during a period that string theorists Gasperini and Veneziano refer to as the "*prehistory* of the universe–starting long before what we have so far been calling time zero–that leads up to the Planckian cosmic embryo." *The Elegant Universe*, supra note 67, at 362. Could this prehistorical period conceivably coincide with or encompass the Omega Point era of a predecessor cosmos and the artifacts of that era, including precisely engineered Calabi-Yau shapes? If so, can we conceive of mechanisms by means of which such engineered shapes could be transmitted by a sufficiently advanced life form across the time zero threshold to the daughter universe? Do the notions put forward by J. Richard Gott III and Li-Xin Li regarding the question of whether the universe can create itself offer any useful clues? See J. Richard Gott III and Li-Xin Li, "Can the Universe Create Itself?" (*http://xxx.lanl.gov/abs/astro-ph/9712344*).
70. Dawkins has criticized Smolin's theory on the ground that it does not, in parsimonious fashion, account for the oddly bio-friendly character of the cosmos. See Richard Dawkins, supra note 16, at 3. The Selfish Biocosm refinement of Smolin's theory outlined here would appear to answer Dawkins' criticism.
71. Wheeler, supra note 6, at 310.

THE INTELLIGENT UNIVERSE

Bibliography

Adams, Douglas. *The Hitchhiker's Guide to the Galaxy.* New York: Harmony, 2004.

Armstrong, Karen *A History of God: The 4,000-Year Quest of Judaism, Christianity and Islam.* New York: Ballantine Books, 1993.

Arnold, Luc F. A. "Transit Lightcurve Signatures of Artificial Objects." Ithaca, N.Y.: Cornell University, March 2005. *http://xxx.lanl.gov/PS_cache/astro-ph/pdf/0503/0503580.pdf.* Accessed by author August 2006.

Baez, J. Online commentary on *The Life of the Cosmos*; *www.aleph.se/Trans/Global/Omega/smolin.txt*, 1998.

Barrow, J. and F. Tipler. *The Anthropic Cosmological Principle.* Oxford, UK: Oxford University Press, 1998.

Barrow, John D. *The World Within the World.* Oxford, UK: Oxford University Press, 1988.

Bedau, Mark, John S. McCaskill, Norman H. Packard, Steen Rasmussen, ed. *Proceedings of the Seventh International Conference on Artificial Life.* Cambridge, Mass: MIT Press, 2000.

Bedau, Mark. "Artificial Life," in *Blackwell Guide to the Philosophy of Computing and Information.* (Luciano Floridi, ed.) Malden, Mass: Blackwell Publishing, 2003.

———. "Four Puzzles About Life." *Artificial Life* magazine, 1998, 125, 137.

———. "How to understand the question 'What is life?'" *Workshop and Tutorial Proceedings, Ninth International Conference on the Simulation and Synthesis of Living Systems (ALife IX).* Boston, Mass., 2004.

Begley, Sharon "Science Journal: Are the universe's traits random or inevitable?" *The Wall Street Journal,* September 16, 2005.

Bjorken, J. "The Classification of Universes," Stanford Linear Accelerator Center, 2004.

Butler, Samuel. "Darwin Among the Machines." *The Notebooks of Samuel Butler.* n.p.: Festing Jones, 1912; quoted in Ray Kurzweil, *The Singularity Is Near: When Humans Transcend Biology* (New York: Viking, 2005), 205.

Carroll, Sean. "The Time Before Time." *Seed* magazine, September 2006. *www.seedmagazine.com/news/2006/08/time_before_time.php?page=all&p=y.* Accessed by author August 2006.

Casti, John. *Complexification.* New York: Harper Collins, 1994.

———. *Paradigms Lost.* New York: Avon, 1990.

Chaisson, Eric. *The Life Era: Cosmic Selection & Conscious Evolution.* New York: Atlantic Monthly Press, 1987.

Chown, Marcus. "Is There Anybody in There?" *New Scientist*, November 26-December 2, 2005, 30.

Cirkovic, Milan M. "Macro-Engineering in the Galactic Context: A New Agenda for Astrobiology." Ithaca, N.Y.: Cornell University. *http://xxx.lanl.gov/ftp/astro-ph/papers/0606/0606102.pdf.* Accessed by author August 2006.

Clarke, Arthur C. *Childhood's End.* New York: Ballantine, 1990.

Cleland, C. "Historical Science, Experimental Science, and the Scientific Method," *Geology* 29, 978-990; 2001.

———. "Historical Science and the Use of Biosignatures," unpublished summary of presentation abstracted in *International Journal of Astrobiology*, 119; 2004.

———. "Methodological and Epistemic Differences Between Historical Science and Experimental Science," *Philosophy of Science* 69, 474-496; 2002.

Crane, L. "On the Role of Intelligent Life in the Evolution of the Universe," unpublished paper, Kansas State University.

———. "Possible Implications of the Quantum Theory of Gravity: An Introduction to the Meduso-Anthropic Principle," *http://xxx.lanl.gov/ftp/hep-th/papers/9402/9402104.gz.*

Crick, Francis. *Life Itself: Its Origin and Nature.* New York: Simon & Schuster, 1981.

Crowe, Michael J. *The Extraterrestrial Life Debate, 1750-1900.* Mineola, N.Y.: Dover, 1999.

Darwin, Charles. *Autobiography (1809-1882).* New York: Norton, 1969.

———. *The Descent of Man.* Princeton, NJ: Princeton University Press, 1981.

———. *The Origin of Species.* New York: Random House, 1993.

Davies, Paul. "E.T. and God," *The Atlantic Monthly*, September 2003.

———. *Edge* magazine interview. *www.edge.org/3rd_culture/davies/davies_p6.html.* Accessed by author August 2006.

———. *The Fifth Miracle.* New York: Simon & Schuster, 1999.

———. "Message for the curious: please phone ET, at home," *The Sydney Morning Herald. www.smh.com.au/articles/2004/08/09/1092022404578.html#.* Accessed by author August 2006.

———. "Transformations in Spirituality and Religion," *When SETI Succeeds: The Impact of High-Information Contact* (Allen Tough, ed.), Foundation for the Future (2000): 51.

Dawkins, Richard. *The Ancestor's Tale.* New York: Houghton Mifflin, 2004.

———. *Climbing Mount Improbable.* New York: Norton, 1997.

———. *The Extended Phenotype.* Oxford: Oxford University Press, 1999.

———. *The God Delusion.* New York: Houghton Mifflin, 2006.

———. Online commentary on "Implications of Natural Selection and The Laws of Physics. *www.edge.org/discourse/index/cgi?OPTION=VIEW&THREAD=richard-dawkins/...smolin.* Accessed by author August 2006.

———. *River Out of Eden.* New York: Basic Books, 1995.

de Duve, Christian. *Vital Dust: Life as a Cosmic Imperative.* New York: Basic Books, 1995.

Dennett, Daniel C. *Breaking the Spell: Religion as a Natural Phenomenon.* New York: Viking, 2006.

———. *Consciousness Explained.* New York: Little, Brown, 1991.

Dick, Steven J. *The Biological Universe.* Cambridge, UK: Cambridge University Press, 1996.

———. "Cosmotheology: The New Universe and Its Theological Implications." Unitarian Universalist sermon. March 18, 2001. *www.uusterling.org/sermons/2001/sermon%202001-03-18.htm.* Accessed by author August 2006.

———. "Cultural evolution, the postbiological universe and SETI," *International Journal of Astrobiology* 2(1) (2003): 65, 71.

———. *Life on Other Worlds: The 20th Century Extraterrestrial Life Debate.* Cambridge, UK: Cambridge University Press, 1998.

Djorgovski, S.G., R. Williams. *Virtual Observatory: From Concept to Implementation.* Ithaca, N.Y.: Cornell University, 2005. *http://xxx.lanl.gov/PS_cache/astroph/pdf/0504/0504006.pdf.* Accessed by author August 2006.

Dobbs, David. "A Revealing Reflection," *Scientific American MIND* (April/May 2006): 22-24.

Dyson, Freeman. *Disturbing the Universe*. New York: Harper & Row, 1979.

———. "Energy in the Universe," Scientific American (September, 1971).

———. *Infinite in All Directions*. New York: Harper Perennial Library, 1988.

———. "Religion from the Outside," *The New York Review of Books* (June 22, 2006): 6.

Edelman, Gerald M. *Bright Air, Brilliant Fire*. New York: Basic Books, 1992.

Epstein, Joshua M. "Agent-Based Computational Models and Generative Social Science," *Complexity* (Vol. 4 No. 5, May/June 1999), 41. *www.setileague.org/iaaseti/abst2002/gardner.pdf.* Accessed by author August 2006.

Fredkin, Edward. "Digital Mechanics: An Informational Process Based on Reversible Universal Cellular Automata." *http://digitalphilosophy.org/dm_paper.htm*. Accessed by author August 2006.

———. "On the Soul." *http://digitalphilosophy.org/on_the_soul.htm*. Accessed by author August 2006.

Gardner, James. "Artificial Exo-Society Modeling: A New Tool for SETI Research," Paper # IAA-02-IAA.9.2 October, 2002. *www.setileague.org/iaaseti/abst2002/gardner.pdf.*

———." Assessing the Robustness of the Emergence of Intelligence: Testing the Selfish Biocosm Hypothesis," *Acta Astronautica*, 48, no. 5-12, 951-955; 2001.

———. "Assessing the Computational Potential of the Eschaton: Testing the Selfish Biocosm Hypothesis," *Journal of the British Interplanetary Society* 55, no. 7/8, 285-288; 2002.

———. *Biocosm—The New Scientific Theory of Evolution: Intelligent Life Is the Architect of the Universe*. Makawao, Maui, Hawaii: Inner Ocean Publishing, 2003.

———. "Coevolution of the Cosmic Past and Future: The Selfish Biocosm as a Closed Timelike Curve," *Complexity* 10, no. 5 (May/June 2005): 14, 17-18.

———. "Genes Beget Memes and Memes Beget Genes: Modeling a New Catalytic Closure," *Complexity* 4, no. 5 (May/June 1999): 22.

———. "Memetic Engineering." *WIRED* magazine, May 1996: 101.

———. "Ninth Circuit's Unpublished Opinions: Denial of Equal Justice?" *American Bar Association Journal* (October, 1975), 1124, 1127.

———. "The physical constants as biosignature: an anthropic retrodiction of the Selfish Biocosm Hypothesis," *International Journal of Astrobiology* 3(3) (2004): 229-236.

———. "The Selfish Biocosm: Complexity as Cosmology," *Complexity* 5, no. 3 (January/February 2000): 34–45.

Gato-Rivera, Beatriz. "Brane Worlds, the Subanthropic Principle and the Undetectability Conjecture." Ithaca, N.Y.: Cornell University Press, 2004. *http://xxx.lanl.gov/PS_cache/physics/pdf/0308/0308078.pdf.* Accessed by author August 2006.

Gee, H. *In Search of Deep Time*. np. The Free Press; 1999.

Gell-Mann, Murray. "What Is Complexity?" *Complexity* magazine 1 (1995): 16, 17.

Gibbs, W. Wayt. *Genetic Programming I, II, II,* and *IV*. Boston: Kluwer Academic Publishers, nd.

———. "Programming With Primordial Ooze." *Scientific American*; *www.genetic-programming.com/published/scientificamerican1096.html.* Accessed by author August 2006.

Giddings, Steve. "The Landscape: A Talk With Leonard Susskind." The Edge Foundation, 2003. *www.edge.org/discourse/landscape.html.* Accessed by author August 2006.

Gingerich, Owen. *The Book Nobody Read: Chasing the Revolutions of Nicolaus Copernicus*. New York: Penguin Books, 2005.

Giraldi, Gilson A., Renato Portugal, and Ricardo N. Thess. "Genetic Algorithms and Quantum Computation." *http://arxiv.org/pdf/cs.NE/0403003*. Accessed by author August 2006.

Gold, Thomas. *The Deep Hot Biosphere*. New York: Copernicus, 2001.

Goldsmith, D. "The Best of All Possible Worlds," *Natural History* 5, no. 6, 44-49; 2004.

Gott, J. Richard and Li-Xin Li. "Can the Universe Create Itself?" Ithaca, N.Y.: Cornell University. *http://xxx.lanl.gov/PS_cache/astro-ph/pdf/9712/9712344.pdf.* Accessed by author August 2006.

Gould, Stephen Jay. *Full House*. New York: Crown, 1996.

———. "Planet of the Bacteria." *www.stephenjaygould.org/library/gould_bacteria.html.* Accessed by author August 2006.

Greene, Brian. *The Elegant Universe: Superstrings, Hidden Dimensions, and the Quest for the Ultimate Theory.* New York: W. W. Norton, 2003.

———. *The Fabric of the Cosmos.* New York: Knopf, 2004.

Grinspoon, David. *Lonely Planets: The Natural Philosophy of Alien Life.* New York: Ecco, 2004.

Hamilton, Craig. "Chasing Immortality:—The Technology of Eternal Life: An Interview with Ray Kurzweil." *What Is Enlightenment?* magazine, September-November, 2005: 61.

———. "Even the Heavens Are Not Immortal—An Alluring Vision of Death: An Interview with Connie Barlow." *What Is Enlightenment?* magazine, September-November 2005: 93-94.

Harrison, E. "Creation and Fitness of the Universe." *Astronomy and Geophysics* 39 (April 1998): 2-27.

———. "The Natural Selection of Universes Containing Intelligent Life." *Quarterly Journal of the Royal Astronomical Society* 36 (1995): 193-203.

Hawking, Stephen, and T. Hertog. "Why Does Inflation Start at the Top of the Hill?" 2002.

Hawking, Stephen, and Roger Penrose. "Quantum Cosmology." *The Nature of Space and Time.* Princeton: Princeton University Press, 1996: 89-90.

Henderson, Lawrence J. *The Fitness of the Environment.* Cambridge, Mass: Harvard University Press, 1913.

Hobbes, Thomas. *Leviathan.* London: Andrew Cooke, 1651.

Hogan, C. "Quarks, Electrons, and Atoms in Closely Related Universes" 2004.

Holland, John H. *Emergence: From Chaos to Order.* New York: Addison Wesley, 1998.

Hoyle, Fred. *The Black Cloud.* New York: New American Library, 1959.

———. "The Universe: Past and Present Reflections." *Annual Review of Astronomy and Astrophysics* 20 (1982): 16-20.

———. "The Universe: Past and Present Reflections." *Engineering & Science* magazine (November, 1981): 8-12, quoted in Owen Gingerich, "Foreword" to Simon Mitton, *Conflict in the Cosmos: Fred Hoyle's Life in* Science. Washington, D.C.: Joseph Henry Press, 2005: xi.

Kassan, Peter. "AI Gone Awry: The Futile Quest for Artificial Intelligence," *Skeptic* 12, no. 2 (2006): 30, 33.

Kauffman, Stuart. *At Home in the Universe.* Oxford: Oxford University Press, 1995.

———. *Investigations.* Oxford: Oxford University Press, 2000.

———*The Origins of Order.* Oxford: Oxford University Press, 1993.

Khoury, J., B.A. Ovrut, N. Seiberg, P. Steinhardt, and N. Turok. 2001 "From Big Crunch to Big Bang"

Koza, John. "Routine Human-Competitive Machine Intelligence by Means of Genetic Programming." Standford University, 2004. *www.genetic-programming.com/cofes2004.pdf.* Accessed by author August 2006.

Krauss, Lawrence M. "Cosmological Antigravity." *Scientific American,* January 1999: 53, 59.

Kurzweil, Ray. *The Age of Spiritual Machines: When Computers Exceed Human Intelligence.* New York: Viking, 1999.

———. "Reinventing Humanity: The Future of Machine-Human Intelligence." *The Futurist* magazine, March-April 2006: 39, 40.

———. *The Singularity Is Near: When Humans Transcend Biology.* New York: Viking, 2005.

Kurzweil, Raymond and Terry Grossman. *Fantastic Voyage: Live Long Enough to Live Forever.* New York: Rodale, 2004.

Linde, A. "The Hard Art of Universe Creation," 2002.

———. "Inflation, Quantum Cosmology and the Anthropic Principle," 2002.

———. "The Self-Reproducing Inflationary Universe," *Scientific American,* 9(20), 98-104; 1998.

Livio, Mario, and Martin J. Rees. "Anthropic Reasoning," *Science* 309, (2005): 1022.

Livio, Mario. "Cosmology and Life," 2003.

Lloyd, Seth. "Life, the Universe, and Everything." *WIRED* magazine, March 2006. *www.wired.com/wired/archive/14.03/play.html?pg=4.* Accessed by author August 2006.

———. *Programming the Universe.* New York: Knopf, 2006.

———. "A theory of quantum gravity based on quantum computation." Ithaca, N.Y.: Cornell University. http://xxx.lanl.gov/PS_cache/quant-ph/pdf/0501/0501135.pdf. Accessed by author August 2006.

———. "Ultimate Physical Limits to Computation," *Nature* 406 (2000): 1047-1054.

Margulis, Lynn, and Dorion Sagan, *What Is Life.*

McFadden, Johnjoe. *Quantum Evolution.* New York: Norton, 2002.

Mead, Walter Russell. "God's Country? Religion and U. S. Foreign Policy." *Foreign Affairs,* September/October 2006: 24, 34-35.

Miller, Kenneth R. *Finding Darwin's God.* New York: Cliff Street Books, 2000.

Minkel, JR. "Hard Landscape: Finding Our Universe in String Theory Appears Impossible," *Scientific American* (2006): 21.

Mitton, Simon. *Conflict in the Cosmos: Fred Hoyle's Life in Science.* Washington, D.C.: Joseph Henry Press, 2005.

Monod, Jacques. *Chance and* Necessity. New York: Knopf, 1971.

Moore, Gordon. "Cramming more components onto integrated circuits." *Electronics,* April 1965. *download.intel.com/research/silicon/moorespaper.pdf.* Accessed by author August 2006.

Morris, Simon Conway *Life's Solution: Inevitable Humans in a Lonely Universe.* Cambridge, UK: Cambridge University Press, 2003.

Mulhall, Douglas. "Are We Guardians, Or Are We Apes Designing Humans?" *Nanotechnology Perceptions: A Review of Ultraprecision Engineering and Nanotechnology* 2, no. 2 (2006). *www.kurzweilai.net/articles/art0673.html?printable=1).* Accessed by author August 2006.

Oldershaw, R. "The new physics: physical or mathematical science?" *American Journal of Physics* 56 (12); 1988.

Olson, Sander. "Interview with Seth Lloyd," November 17, 2002; quoted in Ray Kurzweil, *The Singularity Is Near: When Humans Transcend Biology* (New York: Viking, 2005), 451.

Pagels, Heinz R. *The Cosmic Code.* New York: Bantam, 1983.

———. *Perfect Symmetry.* New York: Bantam, 1986.

Peacocke, Arthur. "The Challenge and Stimulus of the Epic of Evolution to Theology." *Many Worlds: The New Universe, Extraterrestrial Life & the Theological Implications* (Steven Dick, ed.) Philadelphia: Templeton Foundation Press, 2000.

Penrose, Roger. *The Emperor's New Mind.* Oxford, UK: Oxford University Press, 1999.

———. *The Road to Reality: A Complete Guide to the Laws of the Universe.* New York: Knopf, 2004.

———. *Shadows of the Mind: A Search for the Missing Science of Consciousness.* Oxford, UK: Oxford University Press, 1994.

Petit, Charles W. "Touched by nature: Putting evolution to work on the assembly line." *Science,* July 27, 1998; *www.genetic-programming.com/published/usnwr072798.html*). Accessed by author August 2006.

Polanyi, Michael. *Personal Knowledge.* Chicago: University of Chicago Press, 1958.

Raley, Yvonne "Electric Thoughts?" *Scientific American MIND* (2006): 76.

Ramachandran, V. S. "Mirror Neurons and Imitation Learning as the Driving Force Behind 'The Great Leap Forward' in Human Evolution." The Edge Foundation. *www.edge.org/3rd_culture/ramachandran_p1.html.* Accessed by author August 2006.

Rees, Martin J. *Before the Beginning.* New York: Addison Wesley.

———. *Just Six Numbers.* New York: Basic Books, 2000.

———. "Life in Our Universe and Others: A Cosmological Perspective." *Many Worlds: The New Universe, Extraterrestrial Life & the Theological Implications* (Steven Dick, ed.) Philadelphia: Templeton Foundation Press, 2000): 61, 76; quoted in James Gardner, "Dreams of a Cosmic Community," *Science & Spirit. www.science-spirit.org/article_detail.php?article_id=375.* Accessed by author August 2006.

Reynolds, Glenn Harlan. "A Rapture for the Rest of Us." Tech Central Station, 2006. *www.tcsdaily.com/printArticle.aspx?ID=040506B.* Accessed by author August 2006.

Russell, Bertrand. *Study of Mathematics*; quoted in Eugene Wigner, "The Unreasonable Effectiveness of Mathematics in the Natural Sciences." *Communications in Pure and Applied Mathematics* 13, no. 1 (February 1960).

Sakai, Nobuyuki, Ken-ichi Nakao, Hideki Ishihara, and Makoto Kobayashi. "Is it possible to create a universe out of a monopole in the laboratory?" Phys. Rev. D 74, 024026 (2006). *http://xxx.lanl.gov/PS_cache/gr-qc/pdf/0602/0602084.pdf.* Accessed by author August 2006.

Schrödinger, Erwin. *What Is Life?* Cambridge, UK: Cambridge University Press, 1996.

Seager, S., E. L. Turner, J. Schafer, and E. B. Ford. "Vegetation's Red Edge: A Possible Spectroscopic Biosignature of Extraterrestrial Plants." Washington, D.C.: Department of Terrestrial Magnetism, March 2005. *http://arxiv.org/PS_cache/astro-ph/pdf/0503/0503302.pdf.* Accessed by author August 2006.

Shermer, Michael. *How We Believe: Science, Skepticism, and the Search for God.* New York: Henry Holt, 2003.

———. "Shermer's Last Law," *Scientific American* (January 2002): 33.

Shostak, Seth. "Does ET Use Snail Mail?" Mountain View, CA: The SETI Institute. *www.seti.org/site/apps/nl/content2.asp?c=ktJ2J9MMIsE&b=194993&ct=220877.* Accessed by author August 2006.

———. "What Do You Say to an Extraterrestrial?" Mountain View, CA: The SETI Institute. *www.seti.org/site/apps/nl/content2.asp?c=ktJ2J9MMIsE&b=194993&ct=308803.* Accessed by author August 2006.

Smolin, L. *The Life of the Cosmos.* Oxford, UK: Oxford University Press, 1997.

———. "Scientific Alternatives to the Anthropic Principle," 2004.

———. "A Theory of the Whole Universe," *www.edge.org/documents/ThirdCulture/z-Ch.17.html.* Accessed by author August 2006.

"Smolin vs. Susskind: The Anthropic Principle." EDGE: The Third Culture, August 18, 2004, *www.edge.org/3rd_culture/smolin_susskind04/smolin_susskind.html*

Spector, Lee. "And now, digital evolution," *Boston Globe,* August 29, 2005. *http://hampshire.edu/lspector/pubs/digital-evolution.html.* Accessed by author August 2006.

———. *Automatic Quantum Computer Programming: A Genetic Programming Approach.* Boston: Kluwer Academic Publishers, 2004.

Steinhardt, P. and N. Turok. "Cosmic Evolution in a Cyclic Universe," 2001.

Surowiecki, James. *The Wisdom of Crowds.* New York: Anchor Books, 2005.

Susskind, L. "The Anthropic Landscape of String Theory," 2003.

Talcott, Richard. "Is Time On Our Side?" *Astronomy* (February, 2006): 34.

Tarter, Jill S. "The Cosmic Haystack Is Large." *Skeptical Inquirer* (May/June 2006): 31. *www.csicop.org/si/2006-03/cosmos.html.*

———. "SETI and the Religions of the Universe." *Many Worlds: The New Universe, Extraterrestrial Life & the Theological Implications.* (Steven Dick, ed.) Philadelphia: Templeton Foundation Press, 2000.

Tipler, Frank. *The Physics of Immortality.* New York: Doubleday, 1994.

Tough, Allen. "How to Achieve Contact: Five Promising Strategies." *http://members.aol.com/allentough/strategies.html.* Accessed by author August 2006.

Ulum, Stanislaw. "Tribute to John Neumann." *Bulletin of the American Mathematical Society* (1958).

Vakoch, Douglas A. "Altruism: A Scientific Perspective." *Science & Spirit* (September/October, 2001). *www.science-spirit.org/article_detail.php?article_id=237.* Accessed by author August 2006.

———. "Universal Translator Might be Needed to Understand ET." SETI Institute, Jan 2005. *www.seti.org/site/apps/nl/content2.asp?c=ktJ2J9MMIsE&b=194993&ct=363416.* Accessed by author August 2006.

Vilenkin, A. "Anthropic Predictions: The Case of the Cosmological Constant," 2004.

Vinge, Vernor. "The Coming Technological Singularity." San Pedro, Calif.: Accelerate Studies Foundation. *www.accelerating.org/articles/comingtechsingularity.html.* Accessed by author August 2006.

von Neumann, J. "On the General and Logical Theory of Automata," 1948.

Watson, James D. *DNA: The Secret of Life.* New York: Knopf, 2003.

Weinberg, Steven. "A Designer Universe?" *The New York Review of Books*, 1999.

———. *The First Three Minutes*. New York: Basic Books, 1977.

Wheeler, John A. "Foreword" in *The Anthropic Cosmological Principle*, by John D. Barrow and Frank J. Tipler. Oxford: Oxford University Press, 1988.

———. *Geons, Black Holes and Quantum Foam*. New York: Norton, 1998.

———. *At Home in the Universe*. Woodbury, Conn.: AIP Press, 1996.

Whitehead, Alfred North. *Science and the Modern World*. New York: Free Press, 1967.

Wigner, Eugene. "The Unreasonable Effectiveness of Mathematics in the Natural Sciences," *Communications in Pure and Applied Mathematics* 13, no. 1 (February 1960).

Wilson, E. O. *Consilience: The Unity of Knowledge*. New York: Vintage, 1999.

———. "Scientists, Scholars, Knaves and Fools," *American Scientist* 86, 6-7, 1998.

Woit, Peter. *Not Even Wrong: The Failure of String Theory and the Search for Unity in Physical Law*. New York: Basic Books, 2006.

Wolfram, Stephen. *A New Kind of Science*. Champaign, Ill.: Wolfram Media, 2002

Wright, Robert. "Did the Universe Just Happen?" *The Atlantic* monthly (April 1988): 29-44. *http://digitalphysics.org/Publications/Wri88a/html*. Accessed by author August 2006.

Yourgrau, Palle. *A World Without Time: The Forgotten Legacy of Gödel and Einstein*. New York: Basic Books, 2005.

The Intelligent Universe

Index

D

E

F

G

N

O

P

Q

R

S

T

About the Author

JAMES GARDNER

The author of *The Intelligent Universe: AI, ET, and the Emerging Mind of the Cosmos*, James Gardner, is a widely published science essayist and complexity theorist. His science essays and peer-reviewed scientific papers have appeared in *Nature Biotechnology*, *WIRED*, *World Link* (the magazine of the World Economic Forum), the *Wall Street Journal*, *Complexity* (the scientific journal of the Santa Fe Institute), *Acta Astronautica* (the scientific journal of the International Academy of Astronautics), the *International Journal of Astrobiology*, and the *Journal of the British Interplanetary Society*. His first book, *Biocosm*, was selected by Amazon.com's editors as one of the 10 best science books of 2003; he has been featured in major articles in *Time* magazine, *U. S. News & World Report*, *Harper's*, *National Geographic*, and *What Is Enlightenment?* magazine; and he has received outspoken praise from many prominent scientists, including UK Astronomer Royal Sir Martin Rees and astrophysicist Paul Davies.

Gardner has been a featured speaker at major scientific institutions and conferences including the Hayden Planetarium in New York City, the Adler Planetarium in Chicago, the World Space Congress in Houston, the International Astronautical Congress in Rio de Janeiro, the 7th International Artificial Life Conference in Portland, Oregon, and the Chabot Space and Science Center in the San Francisco Bay Area. He currently serves as the principal freelance reviewer of popular science books for *The Sunday Oregonian*.

Gardner is a former Oregon State Senator and a partner in the Portland law and lobbying firm of Gardner & Gardner. He is a graduate of Yale College and the Yale Law School, where he was Article Editor of the *Yale Law Journal*. Following law school, he served as a law clerk for Judge Alfred T. Goodwin on the U. S. Court of Appeals for the Ninth Circuit and for Justice Potter Stewart on the United States Supreme Court.